T0392750

FOOD, GASTRONOMY, SUSTAINABILITY, AND SOCIAL AND CULTURAL DEVELOPMENT

Food, Gastronomy, Sustainability, and Social and Cultural Development
Cross-disciplinary Perspectives

Edited by

F. XAVIER MEDINA
Unesco Chair on Food, Culture and Development, Faculty of Health Sciences, Universitat Oberta de Catalunya/Open University of Catalonia, Barcelona, Spain; International Commission on the Anthropology of Food and Nutrition (ICAF), Barcelona, Spain

DAVID CONDE-CABALLERO
Department of Nursing, Faculty of Nursing and Occupational Therapy, University of Extremadura, Caceres, Spain; International Commission on the Anthropology of Food (ICAF), Caceres, Spain

LORENZO MARIANO-JUÁREZ
Department of Nursing, Faculty of Nursing and Occupational Therapy, University of Extremadura, Caceres, Spain; International Commission on the Anthropology of Food (ICAF), Caceres, Spain

ACADEMIC PRESS
An imprint of Elsevier

Academic Press is an imprint of Elsevier
125 London Wall, London EC2Y 5AS, United Kingdom
525 B Street, Suite 1650, San Diego, CA 92101, United States
50 Hampshire Street, 5th Floor, Cambridge, MA 02139, United States
The Boulevard, Langford Lane, Kidlington, Oxford OX5 1GB, United Kingdom

Copyright © 2023 Elsevier Inc. All rights reserved.

No part of this publication may be reproduced or transmitted in any form or by any means, electronic or mechanical, including photocopying, recording, or any information storage and retrieval system, without permission in writing from the publisher. Details on how to seek permission, further information about the Publisher's permissions policies and our arrangements with organizations such as the Copyright Clearance Center and the Copyright Licensing Agency, can be found at our website: www.elsevier.com/permissions.

This book and the individual contributions contained in it are protected under copyright by the Publisher (other than as may be noted herein).

Notices

Knowledge and best practice in this field are constantly changing. As new research and experience broaden our understanding, changes in research methods, professional practices, or medical treatment may become necessary.

Practitioners and researchers must always rely on their own experience and knowledge in evaluating and using any information, methods, compounds, or experiments described herein. In using such information or methods they should be mindful of their own safety and the safety of others, including parties for whom they have a professional responsibility.

To the fullest extent of the law, neither the Publisher nor the authors, contributors, or editors, assume any liability for any injury and/or damage to persons or property as a matter of products liability, negligence or otherwise, or from any use or operation of any methods, products, instructions, or ideas contained in the material herein.

ISBN: 978-0-323-95993-3

> For Information on all Academic Press publications
> visit our website at https://www.elsevier.com/books-and-journals

Publisher: Jonathan Simpson
Acquisitions Editor: Nicole Denis
Editorial Project Manager: Emerald Li
Production Project Manager: Bharatwaj Varatharajan
Cover Designer: Victoria Pearson

Typeset by MPS Limited, Chennai, India

Contents

List of contributors	*xi*

1. Gastronomy, sustainability, culture. An introduction to contemporary debates

1

Lorenzo Mariano-Juárez, David Conde-Caballero and F. Xavier Medina

Sustainability, food, gastronomy ... where are we going?	2
About this book: order and content	5
References	13

2. Gastronomy, seasonality, localism, and sustainability: lessons from a long memory, a pandemic, and a current war

15

Helen Macbeth

Introduction	15
World war, food, nutrition, and allotments in Britain	18
Seasonality and loss of seasonality	20
Localism	23
Discussion	26
Conclusion and aspirations for the future	27
References	28

3. Why is environmental sustainability not possible without social sustainability? Food, sustainable diets, and sociocultural perspectives in the Mediterranean area

31

F. Xavier Medina

Introduction	31
Addressing social sustainability	33
The Mediterranean diet as a sustainable diet	34
Sustainability and sociocultural aspects: the Mediterranean area as a context	36
Conclusion	43
References	44

v

4. Sustainability at the menu of Malaysian, Indonesian, and Singaporean top chefs — 49

Jean-Pierre Poulain, Siti Ramadhaniatun Binti Ismail and Frederic Cerchi

Emergence of awareness	49
Socioeconomic contexts	51
Methodology	54
Results	58
Discussion and conclusion	76
References	78
Further reading	80

5. Sustainable diets: intergenerational challenges from the Global South — 83

Mariana Hase Ueta

Introduction	83
Meat: the enemy of the environment	84
China and Brazil: the taste for meat	86
Different generations sitting at the same table	89
Conclusion	91
References	92

6. Exploring tourists and local consumers' attitudes on service automation in restaurant industry: the Spanish fast-food experience — 95

Nela Filimon and Francesc Fusté-Forné

Introduction	95
Service robots in tourism: the potential for gastronomy experiences	96
A definition of service robots	97
Automation in food experiences	98
The morphology of robots	99
Human-robot interactions in gastronomic experiences	100
Service robots and consumer attitudes: recent data and trends	101
Data and method	104
Results and discussion	104
Conclusions	110
References	111

Contents   vii

7. Sustainable restaurants in Barcelona (Spain): identity and sustainability in local cuisine   117

Manuela Alvarenga Nascimento

Introduction	117
Sustainable restaurant: the chef as a key social actor	120
Methodology	121
Catalan cuisine: cultural identity and sustainability	123
Seasonality, product awareness, and cooperation with local producers	126
Barcelona as a cosmopolitan city: giving a new meaning to Catalan cuisine	128
Conclusions	130
References	131

8. The cuisine of the new spirit of capitalism: Noma considerations regarding the value of the authentic and other orders of worth   133

Joan Frigolé

Introduction	133
Theoretical approach and main concepts	134
Noma: a model of gastronomic excellence	139
The idea of the test	139
The domain of the wild	140
The wild as an icon of the indigenous	143
The architectural or institutional context	144
Labor relations	144
Cooks and customers without intermediaries	146
A brief contextualization of Noma	146
Acknowledgments	149
References	150

9. Food on wheels: culinary paths toward sustainable lives for migrants in Germany   151

Edda Starck and Raúl Matta

Introduction: forced migration and civic engagement through food	151
Pedaling food bikes to bypass immigration structures	154
Shaping food affects through mobility	159
Conclusion	165
References	166

10. Agri-food routes as tools for sustainable rural development: the case of chili route in Yahualica Denomination of Origin 169

Daniel De Jesús-Contreras, Laura Elena Martínez-Salvador, Emerio Ruvalcaba-Gómez and Frédéric Duhart

Introduction	169
Agri-food routes and sustainable rural development	170
Denomination of Origin Chile Yahualica	172
Gastronomic and food uses of Yahualica chili	175
Tourist and gastronomic activations of Yahualica chili	176
The Yahualica chili route and sustainable rural development	178
Conclusions	181
References	182

11. Gastronomic tourism and alternative food networks: a contribution to the Agenda 2030 185

María del Pilar Leal-Londoño

Introduction	185
Gastronomic tourism and alternative food networks	186
Gastronomic tourism and the Agenda 2030	189
The Pyrenees tourism brand: a case study	190
Discussion and conclusions	194
References	197

12. Gastronomic tradition, sustainability, and development: an ethnographic perspective of gastronomy in Las Hurdes (Extremadura, Spain) 201

David Conde-Caballero, Borja Rivero Jiménez and Lorenzo Mariano-Juárez

Introduction: the construction of the "black legend" of Las Hurdes	201
Methods	205
Abundance versus the narrative of deprivation	206
Exquisiteness, sustainability, and new gastronomy	208
Final considerations	211
References	212

Contents ix

13. Eating and thinking at the same time: food consumption and sustainability in Lugo (Galicia, Spain)   **215**

Elena Freire-Paz

Introduction	215
Devouring nature without feeding people	217
From global to local: food consumption in the city of Lugo	221
One offer that expands against another that contracts?	229
Conclusions: the future was yesterday	233
References	234
Further reading	235

14. The Organic Market of Mazatlan (Sinaloa, Mexico). Paradoxes of food supply, tourism, and migration   **239**

José Antonio Vázquez-Medina, Erika Cruz-Coria, Elizabeth Olmos-Martínez and Mónica Velarde-Valdez

Introduction	239
Conceptual approaches to the configuration of new consumers in the framework of migration and tourism	243
New markets, new consumers, and the generation of emerging food supply spaces in Mazatlan	245
Between the offer, the restriction, and the added value of organic products in the Mazatlán Organic Market	247
From culinary cosmopolitanism to agriculture on demand: new products and culinary information flows in the Mazatlán Organic Market	249
Final considerations	251
References	251

Index   *255*

List of contributors

Frederic Cerchi
Taylor's Culinary Institute, Kuala Lumpur, Malaysia

David Conde-Caballero
Department of Nursing, Faculty of Nursing and Occupational Therapy, University of Extremadura, Caceres, Spain; International Commission on the Anthropology of Food (ICAF), Caceres, Spain

Erika Cruz-Coria
Universidad Autónoma de Occidente, Regional Unity of Mazatlan, Mazatlan, Mexico

Daniel De Jesús-Contreras
Centro Universitario Temascaltepec. Universidad Autónoma del Estado de México, Toluca, Mexico

Frédéric Duhart
Sigmund Freud University, París, Francia

Nela Filimon
Department of Business, University of Girona, Girona, Spain

Elena Freire-Paz
Universidade de Santiago de Compostela, Lugo, España

Joan Frigolé
Department of Social Anthropology, Faculty of Geography and History, Universitat de Barcelona, Spain

Francesc Fusté-Forné
Department of Business, University of Girona, Girona, Spain

Mariana Hase Ueta
Technische Universität Dresden, Dresden, Germany

Siti Ramadhaniatun Binti Ismail
School of Food Studies & Gastronomy, Taylor's University, Kuala Lumpur, Malaysia

María del Pilar Leal-Londoño
CETT- Barcelona School of Tourism, Hospitality & Gastronomy, Universidad de Barcelona, Barcelona, Spain

Helen Macbeth
Anthropology Department, Oxford Brookes University, England, United Kingdom

Lorenzo Mariano-Juárez
Department of Nursing, Faculty of Nursing and Occupational Therapy, University of Extremadura, Caceres, Spain; International Commission on the Anthropology of Food (ICAF), Caceres, Spain

Laura Elena Martínez-Salvador
Instituto de Investigaciones Sociales. Universidad Nacional Autónoma de México, Ciudad de México, Mexico

Raúl Matta
Institute of Cultural Anthropology/European Ethnology, University of Göttingen, Göttingen, Germany

F. Xavier Medina
Unesco Chair on Food, Culture and Development, Faculty of Health Sciences, Universitat Oberta de Catalunya/Open University of Catalonia, Barcelona, Spain; International Commission on the Anthropology of Food and Nutrition (ICAF), Barcelona, Spain

Manuela Alvarenga Nascimento
Department of Social Anthropology, Universidade Federal de Goiás, Goiânia, Brazil

Elizabeth Olmos-Martínez
Universidad Autónoma de Occidente, Regional Unity of Mazatlan, Mazatlan, Mexico

Jean-Pierre Poulain
Faculty of Social Sciences, Taylor's University, Kuala Lumpur, Malaysia; Institut Supérieur du Tourisme, de l'Hôtellerie et de l'Alimentation (ISTHIA), Jean Jaurès University, Toulouse, France

Borja Rivero Jiménez
Department of Business Management and Sociology, University of Extremadura, Cáceres, Spain

Emerio Ruvalcaba-Gómez
Centro Agroalimentario CAXCAN, Jalisco, Mexico

Edda Starck
Institute of Cultural Anthropology/European Ethnology, University of Göttingen, Göttingen, Germany

José Antonio Vázquez-Medina
Universidad Autónoma de Occidente, Regional Unity of Mazatlan, Mazatlan, Mexico

Mónica Velarde-Valdez
Universidad Autónoma de Occidente, Regional Unity of Mazatlan, Mazatlan, Mexico

CHAPTER ONE

Gastronomy, sustainability, culture. An introduction to contemporary debates ☆

Lorenzo Mariano-Juárez[1,2], David Conde-Caballero[1,2] and F. Xavier Medina[3,4]
[1]Department of Nursing, Faculty of Nursing and Occupational Therapy, University of Extremadura, Caceres, Spain
[2]International Commission on the Anthropology of Food (ICAF), Caceres, Spain
[3]Unesco Chair on Food, Culture and Development, Faculty of Health Sciences, Universitat Oberta de Catalunya/Open University of Catalonia, Barcelona, Spain
[4]International Commission on the Anthropology of Food and Nutrition (ICAF), Barcelona, Spain

In the context of discussions on food, and more specifically on gastronomy, a few issues resonate as reflections on sustainability. The present scenario underscores the threats of the days to come. The consequences of climate change or armed conflicts, such as the one in Ukraine, have an impact on both economies and agricultural production chains. A part of the population is beginning to realize that natural resources, and also cultural resources, cannot be taken as goods for their own consumption, and that there is a duty to preserve them for those who will come after them. These reflections have been included in the discussion on the role that gastronomy and its actors should play in pursuing sustainable models of being in the world. The food chain, from production to consumption and, beyond that, to recycling or composting and the return to production, affects the environment, the social and territorial balance, and has an impact on our most local and immediate environment as well as globally.

It is therefore not surprising that gastronomy and the gastronomic industry have undertaken processes of transformation toward environmentally friendly practices, including investment in renewable energy sources and the use of sustainable materials. But also of concern are all those aspects that surround food and those that go far beyond nutrition: from ethics to esthetics, from the origin to the recycling of resources, from productivity and economic

---

☆ This book is the result of a joint academic project between the International Commission on the Anthropology of Food and Nutrition (ICAF) and the UNESCO Chair on Food, Culture and Development of the Universitat Oberta de Catalunya (UOC), in collaboration with research groups GISCSA-Unex and FoodLab-UOC. The book has been anonymously peer reviewed. Those reviews are deposited at the UOC's UNESCO Chair on Food, Culture and Development.

benefit to gender equity and family reconciliation. And all those elements, in a context such as the one described above, production spaces, distribution processes, or restaurants themselves, can become key instruments for working and recreating these discourses and transmitting them to society in general.

This book aims to offer the reader an approach, through different case studies, to various topics in which gastronomy and sustainability are intertwined. We seek to analyze the problems that relate gastronomy—understood here in a broad sense of the concept and following the food chain, from food production to the reuse of waste—food sustainability—as a sine qua non element of local and global analysis and an unavoidable objective of any future development scenario. And culture—understood in its most anthropological sense, as a human creation from a broad and integrated approach. And all of this while also integrating interdisciplinary perspectives.

Sustainability, food, gastronomy ... where are we going?

The incorporation of sustainability issues into food and dietary patterns has been increasingly discussed since the 1980s, with the dual aim of making the latter healthier for both consumers and the environment. Authors like Gössling and Hall (2013) call it a new culinary system that emerged from food citizenship and social justice in the food supply chain. Gussow and Clancy (1986) suggested the term "sustainable diet" to describe a diet primarily based on foods chosen not only in relation to health, but also in relation to sustainability. They concluded that consumers should, where possible, buy locally produced food, making it less expensive in terms of energy (due to the minimization of transport) while supporting local and regional agriculture and livestock farming. In this sense, the concept of sustainable diets recognizes the interdependencies of food production and consumption with dietary requirements and nutritional recommendations and, at the same time, reaffirms the notion that human health cannot be isolated from environmental health (Burlingame & Dernini, 2012).

In this way, the incorporation of issues related to sustainability has been increasingly present in recent decades on the international political and social agenda in terms of agri-food, nutrition, and public health. Thus, supranational institutions, such as the Food and Agriculture Organization (FAO-UN), in line with the Sustainable Development Goals promoted by the United Nations, point out that:

> To be sustainable, agriculture must meet the needs of present and future generations, while ensuring profitability, environmental health, and social and

economic equity. Sustainable food and agriculture contribute to all four pillars of food security — availability, access, utilization and stability — and to all three dimensions of sustainability — environmental, social and economic.[1]

Thus, one of today's main concerns seems to be to conserve natural resources for future generations and, at the same time, to provide sufficient food, in quantity and quality, to meet the nutritional needs of a growing world population. In this regard:

Sustainable, ecosystem-specific diets are just one practical way of applying sustainability to food security and nutrition. In this overall context, sustainability becomes the long-term component of all levels and dimensions of food security, as well as a well-established and accepted determinant of a nation's health and well-being. (Dernini et al., 2017: 1)

More and more, from an environmental point of view, aspects such as local production and consumption, reducing the use of fertilizers, and limiting (or eliminating) genetically modified varieties are being advocated. Authors such as Hailweil (2014) criticize the distances that food has to travel between its origin and its points of consumption, as well as the disconnection between producers and consumers and the excess waste and wasteful use of food. Sustainability in the context of food is linked to slow food, Km 0, zero waste, and local and proximity food production. All these aspects have, for their part, been taken up in the gastronomic field, which, at some of its levels, is trying to modify its guidelines for action along these lines.

However, it should be noted that, as in other fields and different points of attention, and as we will also see in more detail below, cultural aspects have usually been neglected, being seen only as subordinate or, hopefully, complementary to other more important aspects. As an illustration, policymakers have been committed to models linked to environmental sustainability and, very clearly, to the implementation of those aspects related to public health; however, only a simple cultural "acceptability" of the foods present in a diet (understood in this sense as a food system) is mentioned, and not an internal coherence within a cultural system. ¿How is it possible to address the sustainability of food patterns without incorporating what these patterns mean for individuals or groups? For example, Alba Zaluar's ethnography on family roles in the popular urban classes in Rio de Janeiro reports that "food is one of the main vehicles through which the urban poor think of their condition". Can we think of

---

[1] Source: http://www.fao.org/sustainability/es/. Accessed 13.09.22.

decreasing meat consumption under an urbanite and affluent middle-class rhetoric that is difficult to incorporate into the cultural logics of other groups, where it is shown as an aspirational element or the expression of social ascent?

In recent years we have witnessed a renewed call for the unavoidable incorporation of culture into the sustainability equation. And this has problematized a discourse—that of environmental sustainability—until then considered universal. Should the most vulnerable social groups—or nations—be the ones to change their unsustainable practices if they are their only resource or have only just begun to do so as they are at the beginning of the industrialization process? How do we incorporate the call for sustainable development among underdeveloped populations? This leads us to the idea of dispute that surrounds the discourse around "food sustainability," which has not been adopted in the same way or with the same intensity in different parts of the world. It is a complex process with different specificities in each region. We can assume that there are certain "aspirational" consensuses and spaces of tension and dispute in concrete social and cultural practices.

The work of anthropologists, gastronomes, and other social scientists has addressed the transformations that the rhetoric of sustainability has brought about in the restaurant scene: the architecture of spaces, with the use of natural light or ventilation to improve energy demands, the use of eco-friendly cleaning products or mechanisms to combat food waste are combined with other classic characteristics of sustainability, such as the consumption of local products or the return to the seasonality of menus and their adaptation to the calendar. Oosterver et al. (2010: 27) define sustainable food as having natural characteristics, or rather as one that has not been adulterated and has been produced using natural processes; safety characteristics, such as not containing traces of pesticides and other chemicals; comes from farms that take into account animal welfare; is the result of processes that do not harm the environment.

However, research shows that this is not an eminently firm ground, but rather practices that are subject to tension and dispute. According to Spaargaren et al. (2012), some food transition processes are a work in progress. Works such as Onoja and Achike (2015) show that, for several sub-Saharan African countries, the leasing of large surfaces of land that allow people in the north to be fed year-round impacts food transport and distribution processes, not only with a significant carbon footprint, but also making it difficult for local populations to access food, who in turn may suffer from poorer nutrition than when they had access to that land for subsistence farming.

This disputed, ambivalent context can be seen in many other contexts. For example, in the case of meat consumption, the Report on Sustainable Diets published by the EAT-Lancet Commission (Willett et al., 2019) points out animal products as resource-intensive foods due to the emission of greenhouse gases and water footprint. For part of the population, meat has become a symbol of the environmental enemy. The food industry, gastronomy, and catering have offered options to meet these demands—ethical—animal or environmental, but there have also appeared processes of contestation and resistance from other sectors of the population and the industry. There is also evidence of greenwashing processes in gastronomy and catering (Carruthers et al., 2015), with practices that are sold as sustainable but are in fact sales gimmicks in the contemporary tourism market. It has also been pointed out that the multinational control of the food industry decontextualizes food and turns customers away from any sort of reference to their geographic or social roots (Parrot et al., 2002), which for some scholars provides a partial explanation for many of the recent food security crises. The tension drives responses such as the emergence of Alternative Food Networks (AFNs). Talking about food sustainability, the constant seems to be in dispute.

About this book: order and content

This volume provides an approach to the contemporary debate on gastronomy and sustainability based on reflections, case studies, and ethnographies from different places and contexts, such as Denmark, Mexico, Brazil, China, Germany, UK, Spain, Malaysia, Singapore, Indonesia, and the Mediterranean basin. A total of twelve texts offer their contributions to increase the theoretical baggage and critical analysis that the social sciences can provide in the current context of energy and environmental crises and rising prices in a context that many economists are beginning to call inflationary. For example, the context of production and transport of the food that reaches millions of tables every day, without the consumer being aware of the diaspora that has landed on his or her table, is analyzed. The first of the texts, by **Helen Macbeth**, reflects on categories such as seasonality, localism, or sustainability through an interesting personal-diachronic look at localist practices in food production. She started it with the experiences of growing food in gardens, in allotments, and on any land that could be made available in the UK from the Second World War period as part of the strategy of the then newly created Ministry of Food, which included slogans such as "Dig for Victory." The results of those efforts produced remarkable results: Wartime health surveys of the 1940s show that across all socioeconomic levels, there was better overall nutrition of the UK population during the 1940s than before or after. The text discusses how the renewed interest in growing part of what is consumed can be a key practice in the efforts for sustainable gastronomy. It also proposes an approach to the debates on the impact that the loss of seasonality or local food consumption has on the planet based on examples from different geographies, pointing to the need for political action to reduce food transport in the face of the urgent reality of climate change.

In the following chapter, **F. Xavier Medina** addresses the recent incorporation of the category of "social sustainability" into the theoretical debate based on a critical analysis of the sustainability of the Mediterranean diet. In recent years, we have come to terms with the cost of food production and consumption for the environment and also for public health. The claimed "sustainability" of dietary patterns has emerged over the last decade as a public health challenge requiring holistic approaches. The author stresses the need to incorporate social and cultural contexts into such perspectives. Efforts to achieve sustainability must be socially oriented, assuming that environmental sustainability cannot take place without social sustainability. The Mediterranean diet has a lower environmental impact than other dietary patterns and, therefore, a smaller water footprint, lower greenhouse gas emissions, or lower energy consumption and land use. Following previous discourses on the building of the concept of the Mediterranean diet (Medina, 2019), F. Xavier Medina analyzes the particularities of the different contexts and versions of Mediterranean diets, which is organized as a model of conduct not only in terms of medical or environmental recommendations, but as an intrinsic part of Mediterranean cultures and their heritage. By understanding the Mediterranean diet as a food/culinary system and by focusing on cultural coherence, a significant contribution is made to the greater sustainability of the region.

One part of the book deals with the local restaurant space and its relationship to sustainability discourses and practices. The text by **Jean Pierre**

Poulain, **Siti Ramadhaniatun Binti Ismail**, and **Frederick Cerchi** explores how Malaysian, Indonesian, and Singaporean chefs are going to put sustainability on their menus based on the research they are carrying out with 15 restaurants among the top in fine dining in these countries. Since the 1960s, the theme of sustainability has gradually moved up the academic and then the political agenda. It then spread throughout society, becoming an action guide for certain actors in economic life. The globalization of food and gastronomy has had contrasting effects. The posture of the *nouvelle cuisine de terroir*, which took local food habits as a source of inspiration for fine cuisine, was one of the greatest benefits. It has been shown as the starting point of gastronomic decolonization and gastronomic development in many parts of the world. This was one of the favorable conditions for the process of relocating styles of the gastronomy using ingredients, cooking technics, and cuisines. Awareness of the environmental consequences of food choices will be the trigger for the repositioning of certain chefs. However, the terms will take various forms: new supply policy, revegetation of the bases, waste management, and so on. The text presents a systematized analysis of gastronomic practices in the following categories: Wording of the carte and menus; food sourcing (organic/animal welfare, fish, sustainability friendly); vegetablization of the menus in respect of seasons (no season, chef-mentioned seasonality); the chef's values, and localization of food.

Mariana Hase Ueta's text reminds us that the act of eating, besides nutrition, is an act of imagination as it involves memories, social realities, perceptions, and people's aspirations for a better future. Sustainability adds a complex layer to this universe, making people question the environmental impact of their actions, as food production, transportation, and discards have a global ecological footprint. The Global South gathers different development experiences, where the expansion of access to consumption and the nutrition transition in these countries represent a challenge to sustainable development with global impact. In this chapter, she analyzes first-hand ethnographic data on urban food consumption experiences in Brazil and China, focusing on the intergenerational challenges to communicate new environmental concerns and push for a change in diets, where families' memories still play an essential role in constructing the values and practices around food. The paper describes how social mobility affects the ways that people across different generations perceive food as it mobilizes memories of aspirational consumption, where certain dishes embody the meaning of achieving a "better life." Sustainability and the role of gastronomy demand a cross-cultural, intergenerational, and global debate. It shows how urban families in big cities such as Shanghai (China) and Campinas (Brazil) understand and discuss their own food consumption practices, pointing to the challenges of addressing sustainable diets in the Global South.

The chapter by **Nela Filimon** and **Francesc Fusté-Forné** presents a general overview of the initiatives undertaken in the food industry with respect to the introduction of artificial intelligence (AI). Some theoreticians have placed the desire for sustainability in many fields on this type of technological innovation, and this text contributes to that debate in the catering scenario. The authors pay special attention to the restaurant industry, with the aim of offering an updated state-of-art for both the business and the consumers. Evidence found gives support to the hypothesis that many AI elements already implemented by restaurants in the pre-pandemic period have increased during the pandemic and are expected to expand their presence in food tourism further. Evidence on consumers' perception of AI in restaurants and its impact on gastronomic experience does not offer clear-cut results, as both optimistic and skeptical expectations are reflected. Spanish data confirm the trend, showing that the generational effect, among others, plays a role, with younger people more likely to embrace and value the benefits of AI. Overall, both for restaurants and consumers, the positive and negative effects of AI are still to be seen, inviting further research in the field. They analyze attitudes and beliefs toward the presence of AI and service robots in the restaurant industry. Data collected with a quota sampling survey based on gender and age were analyzed with descriptive and multivariate quantitative techniques.

Manuela Alvarenga Nascimento addresses in her chapter the notions of identity and sustainability in the local cuisine of four Slow Food restaurants in Barcelona, Spain. The work, the result of a postdoctoral fellowship at the Food Observatory at the Universitat de Barcelona, proposes an interdisciplinary approach that combines sociological theories on a transition to food sustainability based on the theory of the social actor and anthropological theories that address food identity in multicultural contexts. In this regard, food sustainability is approached from a cultural perspective to analyze how local communities put it into practice. Identity is presented as an essential aspect in this scenario, encouraging social actors to seek paths toward sustainability as a way of conserving the environment and the local food culture. The research shows that in

Barcelona, the perception of sustainability is characterized by the existence of multicultural spaces. Chefs connect with the menus they create while feeling free to exploit a wide range of cultural references and to give a new meaning to their dishes. The study also shows that this notion of sustainability favors sustainable entrepreneurship, respect, and a positive appreciation of suppliers and their products.

Joan Frigolé's chapter immerses us in the exploration of the discursive practices of one of the world's temples of gastronomy, the NOMA restaurant in Copenhagen (Denmark), and how they contribute to the analysis of modern capitalism. Based on an analysis of the concepts of patrimonialization and commodification of the authentic in the general theoretical framework provided by the models of "orders of worth" and "spirit of capitalism," developed respectively by Boltanski and Thévenot, the author discusses six main aspects: the idea of test; the domain of the wild; the wild as an icon of the indigenous; the architectural and institutional context; labor relations; and cooks and customers: without intermediaries. Although much of this discourse and discursive practices seem to be directed toward efforts to achieve sustainable gastronomy—the return to the local, to nature, to particularism, and the negation of globalization—the author concludes that NOMA is one of the varied and differentiated productions essential for the third spirit of capitalism, in terms of both ideology and justification as well as capital accumulation. Varied and differentiated productions are the opposite of standardized and generic productions. These poles of specialization and differentiation are where the productions of patrimonialization and commodification of the authentic are located.

Other chapters delve into the utilitarian uses and management of sustainability. **Edda Starck** and **Raúl Matta's** chapter takes us into the culinary and gastronomic spaces of immigration in Germany. The text explores the role of food in the articulation of relations between migrants and host communities in the public space and studies the practices that in turn develop as a response to these reassembled contexts. Food production, cooking, retail, catering, education, and shared commensality have the potential to accomplish a sustainable role for the presents and futures of migrants in host societies. Also, the authors point out that the power of food for community cohesion and positive interethnic relations invoked by initiatives is often simply based on assumptions, ignoring the discrimination and othering that can occur through food. They focus first on REFUEAT, a catering business in which the employees, all Syrian refugees, use bicycle kitchen trailers specially designed to travel on Berlin's roads and set up portable grill stalls at public events and private occasions. The second initiative is Über den Tellerrand (which translates as "[to look] over the edge of your plate," a German expression for open-mindedness), an association whose focus is to build a community among people of different cultural backgrounds regarding two social integration initiatives taking place in Germany. They provide valuable insights that we use to develop a new materialist perspective in which food, mobile materialities, and humans interrelate and respond to the unequal relations that affect the lives of refugees and migrants. Both initiatives provide economic sustainability and opportunities for encounter and community building through contexts and practices that enable migrants to live on their own terms, rather than in dependence on a generous host society. By reducing economic dependency on institutional support, they help migrants to regain autonomy over their lives. By bringing together migrants and locals on equal terms and in a way that highlights the expertise of the migrants, they furthermore work to deemphasize the notion of assimilation and instead give priority to the cocreation of social inclusion.

The text of **Daniel De Jesús-Contreras**, **Laura Elena Martínez-Salvador**, **Emerio Ruvalcaba-Gómez**, and **Frédéric Duhart** focuses on the potential of agri-routes for sustainable rural development based on a case study of the Yahualica chili route in Mexico. With an interdisciplinary approach involving anthropology, economics, and rural sciences, they describe how this route emerges as tools of support, visibility, and valorization of both chili itself and the figures of the Denomination of Origin. Within the "new strategies" for rural development, authors find initiatives associated with the activation and socio-cultural valorization of food heritage, especially since rural territories are home to agri-food products capable of boosting the local economy and encouraging collective action for their valorization. From their point of view, agri-food routes are potential tools for sustainable rural development because they value local agri-food production, promote the articulation of different sectors, promote cooperation between actors, and contribute to boosting the local economy. These routes can benefit the entire territory through the enhancement of other products and services and the integration of other economic sectors.

María del Pilar Leal-Londoño's text describes a successful model of gastronomic tourism that aligns with the 2030 agenda by applying the concept of networks and sustainability claims on the supply chain of this type of tourism. The network concept reflects relational ways of thinking, which influences many agri-food studies, providing a dynamic character (Kneafsey et al., 2008). This is an analysis of a geographically isolated area such as the Catalan Pyrenees, addressing the practices based on alternative food networks (AFN), which refer to face-to-face interactions such as farmers' markets, proximity relationships that transport the local and regional identity of a commodity, and the extended relationships that shorten the trading space sometimes like fair trade networks (Whatmore et al., 2003). This case study suggests that in contrast with the conventional food systems, the conventions that characterize the farmers, retailers, and restaurants are strongly linked with green or ecological values. Gastronomic tourism and its AFNs may be seen as a new reconnection among farmers and tourists and allow new interaction spaces and scenarios beyond the conventional food supply system. The author points out that the case study for gastronomic tourism can contribute to sustainable local development through the gastronomic attributes identified; however, an increase in the number of stakeholders participating in the development of gastronomic activities that value the gastronomic heritage of the territory is suggested.

The chapter by **David Conde-Caballero**, **Borja Rivero**, and **Lorenzo Mariano-Juárez** shows how the memory of scarcity in a territory can bring back sustainable food practices based on tradition and serve as an impulse for development. The text is based on fieldwork carried out in the region of Las Hurdes (Extremadura, Spain), which for many years has been the example in Spain of the most extreme hunger and poverty. Different studies have shown how this experience of scarcity continues to condition some of today's eating habits. Based on a qualitative ethnographic study, the authors examine the way in which this past continues to influence food and gastronomic customs and habits. The empirical material collected makes it possible to delimit and relate different categories such as scarcity, tradition, and sustainability, revealing an important change not only in practice, but also in the gastronomy and catering of the area, which has opened up a path for tourist attraction and development that appears to be an increasingly important economic lever in a traditionally poor territory.

The last two chapters offer paradoxical perspectives on the discourses and practices of food sustainability. **Elena Freire** takes us into a very interesting theoretical debate based on the analysis of the production and consumption of organic food in the city of Lugo (Galicia, Spain), in which she opposes the consumption of "traditional" products. In a diachronic global-local analysis, she explores how the processes of agrarian transformation that began with the green revolution and contributed to the dismantling of traditional forms in this region, describes the discursive strategies and performative practices that construct "ecological" food consumption—labeling, colors, packaging, places of sale, price...—in contrast to the cultural system in which traditional products—"home-made," local, seasonal—operate. The current consumption of organically produced food is linked to thought insofar as it responds to a process of awareness, reflection, and assumption of a position not free of ethical values and moral considerations influenced by the more than worrying situation of the environment. Meybeck and Vincent Gitz define the characteristics describing these organic products as "quality" attributes and note that they are all "credential attributes," which cannot be tested by consumers and for which they have to rely on "information." The traditional, on the other hand, is inscribed in well-known biographies of these other products. The author underlines the irony that calls for contemporary sustainability demand a return to a successful production model that was dismantled so that the sustainability of the present and of tomorrow is the right to recover one's own culture that was expelled by global dynamics.

The last of the chapters takes us into an ethnography of the organic market in Mazatlan, Mexico. **José Antonio Vázquez-Medina**, **Erika Cruz-Coria, Elizabeth Olmos-Martínez**, and **Mónica Velarde-Valdez** use the anthropology of food to analyze how social processes of a macro order, such as tourism and migration, are capable of intervening in the phases of the local food system that have to do with provisioning and consumption to make the foodscape more complex. Foodscapes have been modified by the arrival of American tourists and the settlement of their migrants. In the case of Mazatlán, this reconfiguration is present not only in the points of sale of ready-to-eat food, but also in the supply points that offer specialized products for new consumers, made up mostly of Americans and Canadians who are permanent migrants, and/or temporary tourists who have settled in the city. One of the supply centers that confirms the above is the Mazatlán Organic Market (MOM): a mobile market that offers organic products installed temporarily during the winter months in the central area of the city. Although these types of initiatives make food production and supply chains more dynamic, shortening the distance between producers and consumers, it is also true that the segmentation of the market to whom it is directed restricts access to these foods to other sectors of the port's inhabitants due to the sale prices and the formats in which the products offered there are sold. In this way, the MOM materializes some important paradoxes regarding access to fresh products free of agrochemicals by the local population, which have been discussed in studies of the sustainability of contemporary food systems. On the one hand, these types of products enable a new marketing space by dynamizing the supply chain. On the other hand, given the difference in consumer prices for organic versus nonorganic products, the target market is highly segmented toward foreign migrants and tourists with ample purchasing power who demand this type of product during their stay in the port, restricting access to other sectors of the Mazatleca population.

In summary, based on these twelve works located in different areas and contexts, this volume aims not only to contribute to contemporary debates on gastronomy and sustainability, offering examples and empirical materials, but also to offer new epistemological windows, new questions that allow us to increase our knowledge of the relationships established between food, gastronomy, sustainability, and culture.

References

Burlingame, B., & Dernini, S. (Eds.), (2012). *Sustainable diets: Directions and solutions for policy, research and action*. Roma: Food and Agriculture Organization of the United Nations (FAO).

Carruthers, C., Burns, A., & Elliott, G. (2015). *Gastronomic tourism: Development, sustainability and applications—a case study of County Cork, Republic of Ireland. The Routledge handbook of sustainable food and gastronomy* (pp. 360—369). Routledge.

Dernini, S., Berry, E., Serra Majem, L., La Vecchia, C., Capone, R., Medina, F. X., Aranceta, J., Belahsen, R., Burlingame, B., Calabrese, G., Corella, D., Donini, L. M., Meybeck, A., Pekcan, A. G., Piscopo, S., Yngve, A., & Trichopoulou, A. (2017). Med Diet 4.0. The mediterranean diet with four sustainable benefits. *Public Health Nutrition, 20*(7), 1322—1330.

Gössling, S., & Hall, M. (2013). *Sustainable culinary systems: Local foods, innovation, and tourism & hospitality*. London: Routledge.

Gussow, J. D., & Clancy, K. (1986). Dietary guidelines for sustainability. *Journal of Nutritional Education, 18*(1), 1—5.

Hailweil, B. (2014). *Eat here: Reclaiming homegrown pleasures in a global super- market*. Washington, DC: World Watch Institute.

Kneafsey, M., et al. (2008). *Reconnecting consumers, producers and food: Exploring alternatives*. Oxford: Berg Publishers.

Medina, F. X. (2019). From healthy to sustainable: Transforming the concept of the mediterranean diet from health to sustainability through culture. In P. Collinson, I. Young, L. Antal, & H. Macbeth (Eds.), *Food and sustainability in the twenty-first century: Cross-disciplinary perspectives* (pp. 57—69). New York and Oxford: Berghahn.

Onoja, A. O., & Achike, A. I. (2015). Large-scale land acquisitions by foreign investors in West Africa: learning points. In consilience. *The Journal of Sustainable Development*, 14. Available from https://doi.org/10.7916/consilience.v0i14.4676.

Oosterver, P., Guivant, J., & Spaargaren, G. (2010). Alimentos verdes em supermercados globalizados: uma agenda teórico-metodológica. In J. Guivant, G. Spaargaren, & C. Rial (Eds.), *Novas práticas alimentares no mercado global* (pp. 15−57). Florianópolis: Ed. da UFSC. (orgs.).

Parrot, N., Wilson, N., & Murdoch, J. (2002). Spatializing quality: Regional production and the alternative geography of food. *European Urban and Regional, 9*, 241−261.

Spaargaren, G., Oosterveer, P., & Loeber, A. (2012). *Food practices in transition: Changing food consumption, retail and production in the age of reflexive modernity*. New York: Routledge.

Whatmore, S., Stassart, P., & Renting, H. (2003). What's alternative about alternative food networks? *Environment and Planning A, 35*(3), 389−391.

Willett, W., et al. (2019). Food in the anthropocene: The EAT-Lancet Commission on healthy diets from sustainable food systems. *The Lancet, London, 393*(10170), 447−492.

CHAPTER TWO

Gastronomy, seasonality, localism, and sustainability: lessons from a long memory, a pandemic, and a current war

Helen Macbeth
Anthropology Department, Oxford Brookes University, Oxford, United Kingdom

Introduction

Using such words as gastronomy, seasonality, localism, and sustainability in a title might seem rather impressive, but the reality can be phrased so much more simply, as I hope to achieve in this chapter. The issue discussed continues the discussions in chapters in earlier books and conferences organized by the International Commission on the Anthropology of Food and Nutrition (ICAF) (e.g., Lightowler & Macbeth, 2014; Macbeth, 2014). For example, I started my chapter in the book, *Food and Sustainability in the Twenty-first Century* (Collinson et al., 2019) with the question: "Has growing food for the family become an anachronism?" (Macbeth, 2019). This chapter is about how and where food is grown, and I argue that it is beneficial for environmental and economic reasons that, where possible, food consumed is grown and produced locally.

In the current and very urgent environmental state of global warming,[1] as well as in the difficult contemporary economic situation with the prices for bought food and energy suddenly increasing together because of the war in Ukraine,[2] it is perhaps time to reconsider the ways of producing food of earlier generations, when so many grew at least some of their own food. Did your parents, grandparents, or their parents grow any of

---

[1] e.g., https://climate.nasa.gov/evidence/; https://climate.nasa.gov/scientific-consensus/.
[2] e.g., https://www.forbes.com/uk/advisor/personal-finance/2022/02/24/war-in-ukraine-cost-of-living-crisis/.

their family's food? Even if not, many of them will have purchased much of their food locally grown and quite probably produced by someone they knew by name. The way I phrased that put it into the past, but to what extent could a modern pattern of food localism be considered a way to reduce food transport and the energy expenditure needed for *any* form of travel beyond short distances? To reduce the excessive patterns of food transportation today, it is important to distinguish between some naïve ambition to try to "turn the clocks back," and a forward-looking concept that unites only some aspects of a previous way of life with the benefits learned from the agricultural and horticultural sciences of today. This would help to achieve sufficient food output for differing environments and optimum nutritional quality, while reducing food transportation.

I argue that too often food transportation has been incurred for unnecessary gastronomic, rather than good nutritional, or even survival reasons. In this chapter the word gastronomy[3] is limited to consumers' preferences for their food choices, including their pleasures in such foods. The word localism[4] is used for buying and consuming food grown locally either by the consumers themselves or by local producers, thereby greatly reducing food transportation and so the burning of energy fuels. That such transportation has become more economical than purchasing food produced locally is presumably due to mass production somewhere involving large economies of scale, which might also throw some doubt on optimum taste and nutrient values because of methods that must include artificially maintaining freshness. All such issues need to be seriously reviewed with far more concern for the global environment.

At the same time, too many modern workplace occupations, especially in developed economies, involve a relatively sedentary lifestyle. To counteract this, people may choose to go to gyms (Mariano & Medina, in press) or maybe cycle or run along roads within airspace that may be contaminated with car exhausts, brake dust, and tire rubber (Jarjour[5]; Al-Habaibeh[6]). Yet, if even small plots of land can be found, individuals can grow some vegetables and fruit, whether in home gardens or on any land made available locally, for example on local authority land, as in the

---

[3] Gastronomy: "The art and science of good eating" (The Concise Oxford Dictionary of Current English 1944).

[4] Localism: "Favouring what is local" (The Concise Oxford Dictionary of Current English 1944).

[5] https://ehjournal.biomedcentral.com/articles/10.1186/1476-069X-12-14.

[6] https://www.eauc.org.uk/cyclists_on_busy_roads.

Gastronomy, seasonality, localism, and sustainability

Figure 2.1 Allotments in an Oxfordshire village. *Photograph © Helen Macbeth.*

British allotment (Fig. 2.1) systems.[7] The equivalent to allotments exists in many countries, and in the USA, they are called community gardens.[8] Growing food for the family requires a number of hours of healthy outdoor physical effort varying with different conditions and different seasons, and so would be beneficial as exercise as well as putting fresh food on the table. This is not a call for a return to a complete subsistence farming as a way of life, but these activities should be perceived as outside of the individuals' normal economic activities or education, or pursued in retirement. Meanwhile, there are also local producers, who may be small or large enterprises, but they are businesses that primarily serve the locality, and their produce travels little.

Meanwhile, in some less economically developed countries, for example in several sub-Saharan African countries, foreign organizations, often from other continents, lease a great deal of the available land (e.g., Onoja & Achike, 2015) often for cash-cropping food on a large scale, much of it

---

[7] Allotments: Areas of land, frequently owned by local authorities, divided into sections leased to individuals for the growing of vegetables and fruit.

[8] In the USA a close equivalent to British allotments are called "community gardens," but in Britain the phrase "community gardens" is used for areas where a community group cooperatively grows vegetables and fruit.

for food to be grown for export with too little going to the local families, who in turn may suffer from poorer nutrition than when they had access to that land for subsistence farming. As Collier and Dercon (2014) argue for parts of sub-Saharan Africa, more food could be grown for local consumption with appropriate investment (state or otherwise) in the development of local farms (not megafarms!) to produce food for the local population. Such farms may sometimes be in association with some smallholder activities as well, but this concept too is not a return to past, or in some cases current, subsistence farming in such areas. Instead, it would be an updated form of localism.

After a report on how an increased stimulus for all individuals to grow and produce a proportion of their own food benefitted the nutrition of the British population in wartime in the 1940s, this chapter will consider how, in our temperate climates, locally grown fresh fruit and vegetables are naturally seasonal, and yet, supermarkets are able to sell such "fresh" foods all year around but at the environmental and economic costs of transport plus unexplained methods for keeping such perishable foods "fresh." Into these topics are woven perspectives on gastronomy, and from all such perspectives, the advantages of localism are advocated.

World war, food, nutrition, and allotments in Britain

A very significant example of localism, discussed before in ICAF (e.g., Lightowler & Macbeth, 2014), is the need for and success in growing food in gardens, in allotments, and on any land that could be made available in the UK, such as beside railway lines and roads, in city parks, and around national monuments, during the Second World War. My interest in the topic started with reports on the unexpected nutritional well-being of the population of the British Isles during that stressful period. Before that war, Britain had been importing about 50% of all its food (Ministry of Food, 1946), mostly by sea. During the war many merchant ships, despite being accompanied by naval ships in convoys, were sunk by enemy action, and their cargoes were lost. So, producing enough food for the population required a whole new system. Fortunately for Britain, such a system was devised, greatly thanks to two individuals, Lord

Woolton[9] and Jack Drummond,[10] and their combined understanding that the problem had to be perceived and tackled holistically. A new Ministry of Food was set up in 1939 as soon as war was declared, and in association with the Ministry of Agriculture, a new system was devised, which included:

- Food rationing,[11] which distributed food fairly to all, whether poor, rich, or royal;
- Definition of what those rations should be, according to the new science of nutrition;
- Restoring farming to farmland left uncultivated during the previous economic depression;
- Opening up more land for cultivation, often in unexpected places;
- The Women's Land Army[12]
- The government not only took control of all food from the farm or port through production processes to retail distribution, *but also* created radio, film, and magazine propaganda for people to grow their own food. The slogan was "Dig for Victory."

.... And the public did! They grew vegetables in any spare land they could find, as suggested above. Also, individuals raised chickens, had beehives, cooperated with neighbors to raise a pig on kitchen scraps, etc., even in urban settings. What is fascinating is that the wartime health surveys of the 1940s showed that across all socioeconomic levels, there was better overall nutrition of the UK population during the 1940s than before or after (Darke, 1979; Huxley et al., 2000; Sinclair, 1944). As these nutritional results contrasted greatly with other countries across Europe and beyond for the same period, there are lessons to be learned, which I argue are also relevant today to the need to maintain food security while adjusting our lifestyle to reduce the causes of global warming. The word sustainability is much used these days with little attempt to define its meaning, and yet it is important to clarify in each case what variable is to be sustained. In this chapter, sustainability is used to refer to sustaining the world's environment as habitable for all current species, while maintaining food supplies for humanity.

---

[9] https://en.wikipedia.org/wiki/Frederick_Marquis,_1st_Earl_of_Woolton.

[10] https://en.wikipedia.org/wiki/Jack_Drummond.

[11] https://www.iwm.org.uk/history/what-you-need-to-know-about-rationing-in-the-second-world-war.

[12] https://www.iwm.org.uk/history/what-was-the-womens-land-army.

Many, who grew their own vegetables throughout the war, continued to do so throughout the 1950s and beyond, but then gradually, fewer people did. Public parks had been returned to places with footpaths and flowers, with sometimes areas of recreation for a sports field or tennis courts. Land beside roads and railway lines was no longer used for growing food, but the tradition of the British allotments was maintained. However, gradually even these were reduced; for example, in several places the land for allotments, usually publicly owned, was used by the local authority for building new homes. It is also true that for a while the number of those wanting to work an allotment was reducing. However, that has changed more recently, and there has been an increase in the demand for allotments; in many urban areas there are now long waiting lists as people again understand the advantages of healthy food and exercise. In my rural village there is today a very short waiting list to get even a half–allotment (about 25 m by 3.7 m), and our Allotments Association here has started to offer quarter allotments. The rental rate is extremely low, the physical work beneficial, and the homegrown food brought to the kitchen fresh.

Seasonality and loss of seasonality

Having introduced my use of the words gastronomy, localism, and sustainability, I now introduce "seasonality," which I use in respect of seasonal changes in the climate as it affects the ability to grow different crops and, in this way, affects when local-grown vegetables and fruit can be eaten fresh. There is a pleasure which in the wealthier developed economies has been disappearing from many people's lives: it is the pleasure of eating the first fruits of a season. I exemplify this with my own recollection of waiting for the first new potatoes of the year, and then the gastronomic enjoyment of those first homegrown potatoes, boiled and served with a little melted butter and chopped fresh chives. Yet now, one can buy such potatoes as "salad potatoes" or "Charlotte potatoes" all year round. I live in a northern climate, where seasons greatly affect the growing of vegetables, fruit, and other crops. For example, seeds must not be sown outside too early in the year in case the plants show above ground before a late hard frost. Our summers are not too hot, and we usually (but not always) have enough rain in all seasons. I don't know how it is (or was!) in other climates, but *waiting for* the new produce of each season adds to the gastronomic pleasure of the taste of the first of the year, just after being harvested. Now, however, with global warming, whatever the latitude of their locality, people are witnessing changes in their seasonality (e.g., Stainforth et al., 2013). Those who grow plants in outdoor conditions will be most aware of these changes, whether it is in temperature, rainfall, and/or the fierceness or frequency of storms. Slowly, but still, only slowly, this is changing what can be grown and its timing.

However, now food provisioning has changed. Many food shops became supermarkets, and some even became hypermarkets. These supermarkets and hypermarkets got bigger, more impersonal, more international, and more multinational. More and more food is now transported across the globe, or simply across a continent, and then transported along further steps of the delivery processes as food may have to be conveyed from a seaport or airport to huge distribution warehouses; then it often travels again within the country to other warehouses for further sorting before going on to the retail outlet or home delivery. Transportation clearly is an important part of the greenhouse gas (GHG) emissions involved in much of the food provisioning of today, but in studying the full production, packaging, and storage (including refrigeration) of food supply chains, Wakeland et al. (2012) show there are many sources of GHG, including vehicle emissions during the diverse journeys from farm to retailer. Vermeulen et al. (2012) argue that 19 to 29 percent mancaused GHG emissions are due to food systems overall. Meanwhile, the impact of the increased amount of warehousing on the environment is multiple, as warehouses become more technologically advanced with refrigeration, cooling, heating, air conditioning, lighting, and all the fixed and mobile handling equipment (Marchant, 2010; Ries et al., 2017). To these effects on the environment should be added the hectares of green agricultural land that are being lost to warehousing, even the building materials of the structures and the tarmac surface of the huge parking facilities needed for the goods vehicles and containers, and then add in the journeys and parking of the workforce. However, to return to gastronomy, is the taste of this much-traveled food the same as when the food item is local and very fresh? Techniques are used to keep such food so-called "fresh," which purists can detect in the taste, but which most of us have come to ignore until we taste the difference in homegrown vegetables. [*Readers: You very likely know the delights of fresh-caught fish, but if the difference in vegetables means nothing to you, try comparing the taste of ordinary supermarket carrots simply boiled to that of some carrots, homegrown, freshly dug, and then boiled.*]

Yet, some foods now common in our British diet have for many generations been imported from distant countries because they do not grow in our climate, such as most citrus fruits, bananas, pineapples, etc., as well as the beans for our coffee and the leaves for our tea. I presume such foods will continue to travel around the world to satisfy expectations that have long been part of our culture. The same is, of course, true for many foods now in many cultures in other areas around the world. Furthermore, there are the specialty foods of a particular locality, which are much appreciated far beyond that locality and for which the French use the word "terroir;" their export could still be maintained with neutral effects on global warming, so long as the transportation costs include a contribution to an equivalent carbon negative feedback loop. For those who seek and can afford such gastronomic delights, I expect that these specialties will continue to be transported, whatever the cost. Today, the environmental urgency to reduce all transport is more talked about than faced, and food transportation is no exception. With the current explosion in the cost of fuels, it would seem to be the ideal time to face this issue, but instead I hear only discussions about the costs of fuel and transport, and the inflation of food prices that follows. The view is that this will result in an inflationary spiral, rather than in any readjustment in ways of living for the sake of the environment.

However, what of the seasonal foods common in one culture in their correct season for that climate, but now also grown in another climate to be transported across the world to the other hemisphere just to be available "out of season" there? This has caused an interesting debate. When green beans grown in Kenya are flown to a region where they are out of season but still bought, consideration can be given to the difference between GHG emissions incurred by the transport and those incurred by growing such beans in greenhouses out of season local to the consumers. To make a valid comparison in such a case, one needs to review the transport used to fly the beans to the recipient area, e.g., whether loaded onto a scheduled passenger flight going *anyway* or loaded onto a transport plane being part of the cargo causing that transport to fly at all. It is not quite that simple, but that highlights one issue. Nevertheless, there would still be transportation from the airport to warehouses, etc. Weber and Matthews (2008) reported on a detailed analysis in the USA of GHG costs for food transported from a different environment compared with the GHG costs of heating greenhouses for growing those same foods and concluded that transport caused considerably fewer GHG costs to the environment than heating greenhouses. Nevertheless, I argue in this chapter that it would be far better still to keep enjoying one's seasonal foods only during their proper season in one's own locality and not to seek such foods out of season at all, whether imported or grown under glass. Sim et al. (2007) also argue that it is generally preferential on environmental grounds for consumers to choose local produce when in season than imported produce, but, unlike my view, they add that for all-year-round availability, cultivation overseas may have environmental benefits compared with local cultivation out of season that uses energy. Yet, if the gastronomy of today is based on being able to buy any food in any season, is this really *diversity* in gastronomy? Is it not *monotony*, whereas seasons cause genuine gastronomic diversity because of the delights of the first products of a new season?

Localism

So, for seasonality, gastronomy, and at the same time, sustainability, I say look again at localism! Shop in those local markets where many stalls are run by genuinely local producers selling local, in-season foods. To be clear, it is not to avoid supermarkets that I make this suggestion, but to seek out foods in season, which may also be available in local supermarkets, and yet *may* have traveled; so, check the labels. In every country, local markets differ, so the description "local market" can be misleading as, in some such markets, not all the fresh food stalls sell *only* local foods, for example, when in the UK, a stall sells strawberries out of season or Spanish oranges all year round. Or, consider the vegetables on sale in a Parisian "local market" in February, many of which could not have been grown locally at that time of year (Fig. 2.2). In the UK, we also have what are called farmers' markets, where stalls are almost always leased to and run by local producers and some to other local artisans for food products. In our area there is one on the first Saturday morning of each month throughout the year (open and selling from 8 a.m. to 1 p.m.). These local producers set up their stalls at different markets on different

Figure 2.2 Vegetables on a Paris market stall in February. *Photograph © Helen Macbeth.*

days at least an hour before the stall opens, usually very early in the morning, and yet they, or some of them, may still have their farms or smallholdings to work on all year round. What is more, it is fun to patronize these farmers' markets and chat with the local producers. Finally, I should just add that for those with access to rural areas (harder for those who live in cities), there are also opportunities for foraging wild fruit, nuts, herbs, and mushrooms (Fig. 2.3) according to season, and some folks fish and some hunt for meat, where this is allowed.

However, if you cease to buy imports, you may be concerned about another perspective. What about the local welfare of those in the producer areas? What about the local welfare of those small-scale farmers in the exporter nations? Too often that has now become just a romantic notion; look again at what is really happening! In Almeria, for example, there are kilometers of plastic-covered hectares producing vegetables for export, worked by migrant laborers, and adding plastic waste all around and to the sea (Caparrós-Martínez et al., 2020; Castillo-Díaz et al., 2021), while small-scale businesses of local producers fail and local village life declines as younger folk have to move to other areas for employment. So, investment in localism

Gastronomy, seasonality, localism, and sustainability

Figure 2.3 Wild mushrooms from the Cerdanya valley. *Photograph © Helen Macbeth.*

is needed in these places as well, with recent changes not exactly reversed but amended to a viable lifestyle for those in the locality maintained by their local produce with as little food transport as possible. An important change in attitude to economics for the sake of our global environment is now urgently needed, but how can it be achieved politically?

In summary, the amount that food travels today is the antithesis of localism. Many foods, as mentioned above, are transported across the world generally to other ecosystems or simply to other economic systems. In the case of the UK, they arrive at a seaport or airport, but that is not the end of their transport. Whether from such a port or from a national producer, today, quantities of food are transported to warehouses, sorted, and then resent to other warehouses for further sorting until finally transported again to be delivered to the retailer or direct to the consumer. Such multiple layers of transport and warehousing all use energy, much of which will still be derived from fossil fuels for a long time yet. Meanwhile, people publicize the environmental benefits of electric engines for vehicles too strongly without considering that they too use the energy created *somewhere*, and these vehicles still leave brake dust and tire particles, harmful both to the environment and to human health if breathed. Then, there is also all the energy used for making steel and the vehicles and engines themselves. Despite the yearly increase around the world in electricity from solar, wind, tidal, or hydropower sources, burning fossil or other fuels to create power will continue for years to come. So, the economics (presumed to be of scale) of such transport and re-transport surely must be reconsidered when global warming is taken seriously enough to support the life options of future generations on this planet. If one finds that the food from the local producer is more expensive, then it is likely that the environmental costs of the supermarket or hypermarket options have not been set high enough to account for the damage to the environment caused by the production and transport of the distant goods. With the effects of the war in Ukraine, these fuel costs have become far more obvious and expensive. Surely, the option is to reduce the need for them with less food transportation and more localism.

Discussion

Although stimulated by the way that, in the Second World War, a British wartime government responded to the loss of food imports with holistic food policies for the nation, and by the resultant nutritional success, the theme of this chapter has been based on the *current* need to react to the urgent reality of climate change. The emphasis has been on reducing the extensive transportation of food. However, in the UK, in 2020, 46% of the food consumed had been imported from abroad, which by value was at that time 48 billion pounds sterling (Department for Environment, Food and Rural Affairs, 2021). This chapter has only been concerned with food, whereas it is becoming clear that there are so many changes to our contemporary ways of living that must be achieved to attain the almost unattainable reductions in carbon emissions needed to slow down global warming. We *know* so; we *say* so, but do we yet *act* so?

Then in 2020, Covid-19 spread around the world, and localism took on a new meaning for populations in different areas and at times when they suffered "lockdowns," which, for many, led to new patterns of working from home, very limited travel, and new attitudes to shopping either very locally, for example in local open-air markets, or ordering their supermarket groceries online with deliveries to the home. Depending on the efficiency of the routes achieved by the supermarket's delivery vans and the amount of refrigeration, etc., on board (United States Environmental Protection Agency, 2016), such deliveries, especially in rural areas by vans delivering to many homes on a computer-designed route, could cause lower GHG emissions than all the personal trips by car to a supermarket. Of course, if the purchaser can walk or cycle to shop, or even take public transport to that store, that would be more efficient still. As we adjust to the pandemic with our vaccinations and the (so far) less virulent variants of Covid, many have maintained shopping for groceries online, who had not shopped that way before.

Now we have war again in Europe, and although currently no fighting is beyond the borders of Ukraine, the effects are way beyond those borders, as energy costs and food prices have suddenly soared globally: energy because of sanctions against Russian oil and gas; food because Ukraine had produced and exported about a fifth of the world's high-grade wheat and a quarter of the global sunflower seeds for cooking oil. Add to all this how the costs of energy have affected the costs of production and transport of food and its components. So, living costs are rising and are forecast to continue rising for several months. Yet, will people adapt by choosing to change their lifestyles to a lower level of consumption in general and toward localism in particular, or will they seek or strike for more income to afford a lifestyle that, across the populations of developed economies, has become unaffordable for the environment? Whitmarsh et al. (2021) argue that profound changes in behavior are needed in both individual and community consumer actions, and recommend the need for an interdisciplinary approach, whereas Aveerchenkova et al. (2021) emphasize the need for a legal framework to achieve such changes. Here, it is important to note that I deliberately avoid all the political issues of inequality in access to resources, but that does not mean that these should be ignored, because the effects of the above are proportionately much more significant to the poorest globally and in each society (Salm et al., 2021).

Conclusion and aspirations for the future

The main message of this chapter has been the benefits of consuming food grown and produced locally with the hope that this would reduce the use of fossil fuels as energy for transport, for refrigeration while transported, and for all the needs of warehousing. However, it has also suggested the gastronomic delights of eating locally grown fresh fruits and vegetables as they ripen in a new season, rather than the neglect of seasonal changes and the monotony of having such products available all year round in supermarkets because of long-distance transport and/or modern methods of storage, with any loss of flavor due to these.

I fear it may just be the wish of a geriatric that we might all learn lessons from our experiences of the pandemic and from recent dramatic increases in the prices of energy and food, as well as from what a past population could achieve in wartime with reduced food imports, in order for us to achieve styles of living, learned from the past but reinvented for the future with the benefits of modern sciences. What is needed is a renewed attitude toward (probably reduced) standards of living compatible with the urgent need to reduce carbon emissions to a zero balance with natural sources of carbon intake, rather than reacting to and causing monetary inflation so as to maintain our environmentally unaffordable lifestyles. One lesson for sustainability that I propose is to seek what is local and to enjoy the gastronomy of diversity that comes with the different foods that are fresh in the different seasons in one's own locality. Well, why not grow some of it yourself if any soil is available to you, even if only on an urban balcony or rooftop?

References

Aveerchenkova, A., Frankhouser, S., & Finnegan, J. J. (2021). The impact of strategic climate legislation: Evidence from expert interviews on the UK Climate Change Act. *Climate Policy*, *21*(2), 251–263. Available from https://doi.org/10.1080/14693062.2020.1819190, Accessed 23.06.22.

Caparrós-Martínez, J. L., Rueda-López, N., Milán-García, J., & de Pablo Valenciano, J. (2020). Public policies for sustainability and water security: The case of Almeria (Spain). *Global Ecology and Conservation*, *23*, e01037. Available from https://www.sciencedirect.com/science/article/pii/S2351989420302250, Accessed 24.06.22.

Castillo-Díaz, F. J., Belmonte-Ureña, L. J., Camacho-Ferre, F., & Tello-Marquina, J. C. (2021). The management of agriculture plastic waste in the framework of circular economy. Case of the Almeria Greenhouse (Spain). *International Journal of Environmental Research and Public Health*, *18*(22), 12042. Available from https://www.ncbi.nlm.nih.gov/pmc/articles/PMC8625533/, Accessed 23.06.22.

Collier, P., & Dercon, S. (2014). African agriculture in 50 years: Smallholders in a rapidly changing world? *World Development*, *63*, 92–101.

Collinson, P., Young, I., Antal, L., & Macbeth, H. (Eds.), (2019). *Food and sustainability in the twenty-first century: Cross-disciplinary perspectives*. New York and Oxford: Berghahn.

Darke, S. J. (1979). A nutrition policy for Britain. *Journal of Human Nutrition*, *33*, 438–444.

Department for Environment, Food and Rural Affairs. (2021). *Food security report 2021: Theme 2: UK food supply sources*. London: UK Government Official Statistics.

Huxley, R. R., Lloyd, B. B., Goldacre, M., & Neil, H. A. W. (2000). Nutritional research in world war 2: The Oxford nutrition survey and its research potential 50 years later. *British Journal of Nutrition, 84,* 247–251.

Lightowler, H., & Macbeth, H. (2014). Nutrition, food rationing and gardens in the Second World War. In P. Collinson, & H. Macbeth (Eds.), *Food in zones of conflict: Cross-disciplinary perspectives* (pp. 53–64). New York and Oxford: Berghahn.

Macbeth, H. (2014). 'Dig for Victory!' Do lessons from the past suggest a way forward in the present crisis? In I. González-Turmo (Ed.), *Respuestas Alimentarias a la Crisis Económica.* ICAF-eBOOKS, ch.14, online Kindle Edition.

Macbeth, H. (2019). 'Dig for sustainability' in the twenty-first century: Allotments, gardens and television. In P. Collinson, I. Young, L. Antal, & H. Macbeth (Eds.), *Food and sustainability in the twenty-first century: Cross-disciplinary perspectives* (pp. 113–126). New York and Oxford: Berghahn.

Marchant, C. (2010). Reducing the environmental impact of warehousing. In A. McKinnon, S. Cullinane, H. Browne, & A. Whiteng (Eds.), *Green logistics* (pp. 167–192). London and Philadelphia: Kogan Page.

Mariano, L., & Medina, F. X. (In press). Eating pure; Ethnography and food in "Fitness Cultures". In P. Collinson & H. Macbeth (Eds.), *Pure food: Theoretical and cross-cultural perspectives.* New York and Oxford: Berghahn.

Ministry of Food. (1946). *How Britain was Fed in War Time: Food Control 1939–1945.* London: HMSO.

Onoja, A. O., & Achike, A. I. (2015). Large-scale land acquisitions by foreign investors in West Africa: Learning points. *Consilience, The Journal of Sustainable Development, 14*(2), 173–188.

Ries, J. M., Grosse, E. H., & Fichtinger, J. (2017). Environmental impact of warehousing: A scenario analysis for the United States. *International Journal of Production Research, 55* (21), 6485–6499.

Salm, L., Nisbett, N., Cramer, L., Gillespie, S., & Thornton, P. (2021). How climate change interacts with inequity to affect nutrition. *Wiley Interdisciplinary Reviews: Climate Change, 12*(2), e696. Available from https://doi.org/10.1002/wcc.696, Accessed on 23/06/2022.

Sim, S., Barry, M., Clift, R., & Cowell, S. J. (2007). The relative importance of transport in determining an appropriate sustainability strategy for food sourcing: A case study of fresh produce supply chains. *International Journal of Life Cycle Assessment, 12*(6), 433-431.

Sinclair, H. M. (1944). Wartime nutrition in England as a public health problem. *American Journal of Public Health and the Nation's Health, 34*(8), 828–832.

Stainforth, D. A., Chapman, S. C., & Watkins, N. W. (2013). Mapping climate change in European temperature distributions. *Environmental Research Letters, 8*(3), 034031. Available from https://doi.org/10.1088/1748-9326/8/3/034031.

United States Environmental Protection Agency. (2016). What if more people bought groceries online instead of driving to a store? Available at https://nepis.epa.gov/Exe/ZyPdf.cgi?Dockey = P100O0EL.pdf. Accessed 23.06.22.

Vermeulen, S. J., Campbell, B. M., & Ingram, S. I. (2012). Climate change and food systems. *Annual Review of Environment and Resources, 37,* 195–222.

Wakeland, W., Cholette, S., & Venkat, K. (2012). Food transportation issues and recing carbon footprint. In J. I. Boye, & Y. Arcand (Eds.), *Green technologies in food production and processing* (pp. 211–236). New York: Springer, Food Engineering Series.

Weber, C. L., & Matthews, H. S. (2008). Food-miles and the relative climate impacts of food choices in the United States. *Environmental Science and Technology, 42*(10), 3508–3513.

Whitmarsh, L., Poortinga, W., & Capstick, S. (2021). Behaviour change to address climate change. *Current Opinion in Psychology, 42,* 76–81.

CHAPTER THREE

Why is environmental sustainability not possible without social sustainability? Food, sustainable diets, and sociocultural perspectives in the Mediterranean area

F. Xavier Medina[1,2]
[1]Unesco Chair on Food, Culture and Development, Faculty of Health Sciences, Universitat Oberta de Catalunya/Open University of Catalonia, Barcelona, Spain
[2]International Commission on the Anthropology of Food and Nutrition (ICAF), Barcelona, Spain

Introduction

Food systems are currently evolving and transforming at a rapid rate in our societies. In developed countries, plentiful food supplies, greater production, and effective distribution of products at an industrial level have enabled, for around seventy years, continuous access to food products at relatively affordable prices. Unlike any other time in history, we are currently producing more than enough food to meet the needs of the population, and even producing a significant surplus. This surplus production ensures food stability, at least in the most developed countries, which means the general population has access to food products at all times.

However, it is also true that the huge number of processes and transformations that the food industry goes through often generates a disquieting and growing mistrust among consumers (Medina, 2015) Increasingly we turn to explanatory concepts such as "sustainable," "traditional," "local," "bio," "organic," etc., at the same time the ever greater reference to the need for sustainable and ethical food occupies both a reflexive and political place in a growing number of consumers' minds (Boström & Klintman, 2009). In addition to this resignification of food, there are also the nutritional attributes and the constant references to health through food, which have become important markers that are useful not only for characterizing the new (and not so new) consumers but also for understanding the sheer scale of complexity of contemporary food. A scale that not only encompasses production, distribution, and direct consumption, but goes much further to include spaces such as those of gastronomy and catering, and even the reuse and recycling of food waste.

Today there is increasing evidence of the environmental cost of food production and consumption, as well as the cost to public health and, finally, society in general (Bottalico et al., 2016). The demand for "sustainability" of dietary patterns arose in the last decade as a public health challenge (O'Kane, 2012; Buttriss & Riley, 2013) and became part of the international debate on sustainability (Burlingame & Dernini, 2011), food security and nutrition (Berry et al.,2015; Food and Agriculture Organization of the United Nations, 2013a; Food and Agriculture Organization of the United Nations, 2013b; 2013c; Garnett, 2013), and climate change (Food and Agriculture Organization of the United Nations, 2016). The above assertion can even be corroborated within the framework of the guiding principles of the 2030 sustainable development agenda, the goals of which include the eradication of hunger and malnutrition and global health (or planetary health).

This debate has also taken hold in the Mediterranean area. In this area, one of the most important challenges is still food and nutritional security (Food and Agriculture Organization of the United Nations, 2012a; Food and Agriculture Organization of the United Nations, 2012b), and also, increasingly, sustainability in relation to food, or more specifically, *diets* (Burlingame & Dernini, 2011); also, albeit in a secondary and much more peripheral way up to now, the ever-deferred debate linked to food culture and/or heritage (Medina, 2019a).

In this regard, it is important to highlight that to promote a transition toward more sustainable food systems, a holistic approach needs to be developed covering different spheres and areas: agriculture, economy, culture and lifestyles, environment, climate change, nutrition, and health (Bottalico et al., 2016). And this is crucial if we are to develop effective intersectoral policy instruments that enable improved sustainability of diets and food systems.(Adinolfi et al., 2015). In this regard, and in light of the importance of sustainability on the international public agenda and in political discourse, initiatives in this area mostly seem to lean toward one rather restrictive approach, which gives priority to environmental aspects over other areas that need to be considered within the sustainability paradigm.

Given that food should be considered a biopsychosocial phenomenon where nature and culture intersect, this paper proposes a reflection that reconsiders the importance of taking a broader and more comprehensive (and, above all, social) approach when problematizing dietary patterns in terms of sustainability.

Addressing social sustainability

As indicated by Eizenberg and Jabareen (2017), there is still a lack of theoretical and empirical studies on social sustainability. The literature has revealed that the "social" aspect has been integrated very late into the debates on sustainable development and is therefore still one of the least explored aspects.

The authors suggest that the current situation with regard to climate change, evidence of which has been available for some time, places us at a risk scenario that we need to tackle as societies. However, they also state that risk as a concept is part and parcel of sustainability and that, at the same time, these environmental risk conditions and their consequent uncertainties pose serious social, spatial, structural, and physical (we would also add cultural) threats for contemporary human societies and their living spaces.

These two authors affirm, from these initial premises, that within the framework of environmental sustainability, social sustainability is making efforts to address the risk while at the same time addressing existing social concerns, which means that, without socially oriented practices, the efforts to achieve (environmental or any other type of) sustainability will be undermined, as there are too many gaps between practice and theory. From this perspective, they proposed a comprehensive conceptual framework for social sustainability, composed of four interrelated concepts, each one of which incorporates important social aspects: equity (with economic restructuring, parity of participation, and opportunities with significant public participation); Security (with regard to existing risk); responsible production, from both an environmental and social viewpoint (here we would add labor rights, fair wages and salaries, work-life balance, etc.); and spaces and ways of life that promote a sense of belonging, community, security, health and attachment to place, among other environmental objectives.

In this way, social sustainability points toward the vital development of societies and specific social groups, with a focus on coherence, strengthening, cohesion, and stability of the populations of which they are composed. The concept is principally (although not exclusively) applied to social sectors or disadvantaged or socially unprotected populations, seeking responsible and long-lasting management of existing and future resources. This ultimately implies ensuring that human activity does not compromise the environment, favoring the enduring permanence of human lifestyles and cultures. In this way, social sustainability and sustainable development share as a principle the idea that natural resources cannot be used irrationally. This would cause them to be exhausted. And given that human communities depend on access to these resources, guaranteeing their availability is a matter of vital importance.

However, we need to take into account that one type of sustainability (environmental) cannot take place without the other (social). As though we were looking at an image of Maslow's pyramid of needs, we see that human societies need to meet their own needs, in a particular order (following the same visual idea, as we meet our most basic needs, we develop needs and desires further up the pyramid), and all within the framework of their own internal contradictions.

There is therefore a need to broaden the concept of sustainability to make it more holistic, observed as a series of interlinked structures that create the bridges we need between sociocultural and environmental concerns. In this way we recover some of the ideas put forward by Dempsey et al. (2011), which include in their revision of social sustainability fundamental aspects such as lifestyles, traditions, or a sense of community and an attachment to the said community. A similar idea is expressed by other authors such as Rivero et al. (2021) when they include in their analysis the concept of "tradition" linked both to the community and to sustainability itself.

The Mediterranean diet as a sustainable diet

As indicated by Dernini et al. (2016), the incorporation of issues related to sustainability in the international agrifood and nutrition agendas has become increasingly evident during the last decade. However: What is a sustainable diet?

As highlighted by the FAO in today's most widely accepted definition of sustainable diets, defined at its head offices in Rome during the

International Scientific Symposium: *Biodiversity and Sustainable Diets. United against Hunger,* "Sustainable diets are protective and respectful of biodiversity and ecosystems, culturally acceptable, accessible, economically fair and affordable; nutritionally adequate, safe and healthy; while optimizing natural and human resources" (Burlingame & Dernini, 2011).

In this regard, the concept of sustainable diets recognizes the interdependence between food production and consumption and food needs and nutritional recommendations and, at the same time, expresses the notion that the food system (including production, distribution and consumption, health, economic and social, and cultural aspects, among others) cannot operate outside the ecosystem (Dernini et al., 2016).

We also know that interest in the Mediterranean diet as a sustainable dietary pattern has grown in recent decades (Gussow, 1995; Burlingame & Dernini, 2011; Dernini et al., 2016). The notion of the Mediterranean diet has undergone a progressive evolution over the past six decades, from that of a healthy dietary pattern for the heart to the current model of a sustainable diet (Dernini et al., 2016), including culture as an intermediate phase, with the Mediterranean diet being recognized as an "Intangible Cultural Heritage of Humanity" by UNESCO (Medina, 2019b).

In this respect, various authors affirm that the Mediterranean diet, a plant-based diet with low consumption of animal products, has a lower environmental impact than other dietary patterns and, consequently, has a smaller water footprint and lower greenhouse gas emissions compared with other current dietary patterns with lower energy consumption and land use (Gussow, 1995; Burlingame & Dernini, 2011; Heller et al., 2013; Sáez-Almendros et al., 2013; Tilman & Clarck, 2014; International Center for Advanced Mediterranean Agronomic Studies CIHEAM and Food and Agriculture Organization of the United Nations FAO, 2015; Dernini et al., 2016).

However, although some experts also agree that the Mediterranean diet as a model should provide greater environmental benefits in the Mediterranean region, which is characterized by growing water scarcity (Capone et al., 2013; Vanham et al., 2016), it is also the case that Mediterranean diets (in plural, and applied to the different cultural and regional expressions of the said dietary model) are not standardized, and of course are not all equal in terms of environmental sustainability. For example, as indicated by Aboussaleh et al. (2017), countries in the north eastern Mediterranean region have a bigger per annum and per capita water footprint (2279 m3) than the countries of North Africa (1892 m3), the Balkans (1708 m3), and the Middle East (1656 m3).

The urban-rural dichotomy or the socio-economic differences between the countries in the region also play an important role, which needs to be taken into account. The water footprint, for example, varies a great deal between the different countries in the region. Approximately 91% of the region's water footprint is due to the production and consumption of agricultural products and their by-products (Hachem et al., 2016), including production related to biofuels and animal feed. In this way, the increased demand for food has a direct effect on the volume of water used for irrigation.

Following Tomé (2021), we have seen that these processes are notorious in some places that are radically transforming their productive activity and their socio-environmental relationships to satisfy the global demand for products that are at the base of the Mediterranean diet. For example, the medical recommendation of olive oil and its subsequent identification as a fundamental element of the Mediterranean diet, joined with other related factors, has necessitated an increase in oil production. In the Spanish case, this has been achieved in two different ways: by expanding the area dedicated to olive groves and by watering the existing ones (Tomé, 2021). This growth is making an increasing number of Andalusian municipalities depend on single-crop farming, with all the related risks and effects. As Gómez-Limón (2010) already pointed out more than a decade ago, the expansion and intensification of the olive grove have produced negative environmental effects, such as soil erosion, overexploitation of water resources, diffuse contamination of water, loss of biodiversity, and deterioration of traditional agricultural landscapes.

We also have to say that these kinds of problems have become especially evident in recent times, such as the increasingly frequent droughts experienced by Spain and other Mediterranean countries (France, Italy, etc.) already important at the beginning of the year 2022 (Crellin & Harvey, 2022) and whose latest and hardest manifestation, at the time of writing these lines, was during the summer of 2022.

Sustainability and sociocultural aspects: the Mediterranean area as a context

But this sustainability, so widely demanded today, has a social and cultural dimension in addition to its environmental dimension.

The intended lower environmental impact based on aspects such as more seasonal consumption of fresh and local products, or the diversity of local food products, together with other aspects, such as traditional dishes and conviviality, are the cornerstone of the Mediterranean diet heritage conservation acknowledged by UNESCO in 2010.

Until now, the Mediterranean diet has mainly been seen as a model of healthy behavior based on medical recommendations. However, following its recognition as an Intangible Cultural Heritage by UNESCO (2010−13), the Mediterranean diet is increasingly being seen as an intrinsic part of Mediterranean cultures, with the concept being reformulated to treat it as equivalent to a Mediterranean food system or Mediterranean culinary system (González Turmo, 2005; Medina, 2015; Dernini et al., 2016).

In effect, the "Mediterranean Diet" was inscribed on the UNESCO Representative List of the Intangible Cultural Heritage of Humanity as:

> ... a set of skills, knowledge, practices and traditions ranging from the landscape to the table, including the crops, harvesting, fishing, conservation, processing, preparation and, particularly, consumption of food [...]. This unique lifestyle determined by the climate and Mediterranean landscape is also expressed through associated festivities and celebrations. (Mediterranean Diet, 2010).

Furthermore, through its social and cultural functions, the Mediterranean diet embodies landscapes, natural resources, and associated occupations, as well as the fields of health, well-being, creativity, intercultural dialog, and, at the same time, values such as hospitality or communal living, sustainability, and biodiversity (*cf.* Serra-Majem & Medina, 2016).

Following this definition, the Mediterranean diet is a concept that aims to cover biodiversity, sustainability, quality, health, and culture. Safeguarding the Mediterranean diet should therefore be the driver of responsible and sustainable consumption (González Turmo & Medina, 2012). And equally, from a local Mediterranean point of view, and as a local consumption model, Mediterranean food and *diets* could become a sustainable development resource for the Mediterranean area in terms of local consumption (Medina, 2011).

However, these premises often collide with points of view focused on certain partial interests. For example, they may collide with viewpoints centered mainly on health, based on which the Mediterranean diet is seen more as a list of healthy ingredients than a sociocultural and integrated food/culinary system that goes way beyond nutrition. From a health perspective, the Mediterranean diet can also be seen as an artificially exportable article that *may improve* the health of different societies around the world. In this regard, as indicated by Trichopoulou and his collaborators:

(...) People (at a global level) could try to adjust their diets to the principles of the traditional Mediterranean diet (...) After all, this diet not only promotes health, as indicated by the overwhelming evidence, it is also delicious, as has been recognized by the many people who have tried it in its different varieties. (Trichopoulou et al., 2014: 112–113).

In this same article, he also refers to the concept of a *modernized Mediterranean diet*, which takes into account only the variables related to health, independent of other aspects (cultural, geographical, production and consumption, etc.) considered to be secondary and even dispensable:

The modernized concept of the Mediterranean diet opens the way for a scientifically established protective dietary pattern that could exist independently of geography, climate, and Mediterranean cultures. Future research, for example, pointing to the benefits of a modern Mediterranean diet in an observational epidemiological study, should incorporate this new knowledge (Trichopoulou et al., 2014: 112–113).

Along these same lines, Lacatsu et al. (2019) say that:

In addition to the cardiovascular, metabolic, cognitive and possible antineoplastic benefits, the Mediterranean diet could be associated with good adherence scores in some populations outside the Mediterranean, and with a better quality of life, and nowadays it is recommended by a large majority of professionals throughout the world. At the same time, the erosion of traditions and cultures in populations neighboring the Mediterranean means that its domestic survival is becoming increasingly difficult. Efforts are needed in these apparently disjunctive directions of Mediterranean and non-Mediterranean populations, so that humankind can benefit from this complex web of habits associated with food which began in ancient times as a mixture of lifestyle, religion and lay culture and which ended up becoming an emerging medical prescription for health. (Lacatsu et al., 2019).

After reading these paragraphs, it seems that we are perhaps forgetting that collective or individual dietary habits are always a sociocultural construction and that it is difficult to impose them or implant them artificially. As González Turmo indicates in this respect:

Food choices, despite depending on the person who does the shopping and cooking in each household, are rarely random, even when it might seem to be one of the human activities where personal opinions and tastes play a greater role. The attitudes and behaviors mentioned are, in fact, conditioned by factors (from cultural experience) that are beyond the control of each of us as individuals. (González Turmo, 1996: 13).

We tend to forget that diets go far beyond the nutritional importance of what we consume; that we do not only consume fats, proteins, or vitamins that have a beneficial or negative effect on our organism, but rather, we do so following a series of determined everyday rhythms, particular guidelines for the different foods we consume, within an extensive plurality of cuisines and culinary knowledge, adopting one or another kind of conviviality, and ways of eating together (Medina, 2021). In short, lifestyles that broadly influence what we eat, when, where, and how.

As pointed out by French anthropologist Hubert (1998: 157) more than two decades ago, we are presented with a scientific construction of a representation that, despite offering us a healthy dietary model, which is undoubtedly beneficial, unnecessarily tries to legitimize itself on the basis of a supposed traditionality and a totally conscious choice on the part of those who form or do not form part of this ideal model. We could also add that:

No food can be included or maintained artificially over time (in a culinary system) if it does not fit in with the local cultural habits and processes. This underscores the value of a bio-social, cross-cultural and interdisciplinary approach in dietary studies. (Macbeth & Medina, 2020: 141).

In this regard, and picking up the initial thread of this chapter, we see that no environmental sustainability (or public health) concept can be understood without taking into account the social dimension. Equally, as pointed out by Moser (2001), sustainability should be seen as a development model that is ecologically viable but also socially just and economically competitive, within which all activity should be sustainable by definition.

It should be said that these views do not take into account, for example, aspects such as the sustainability or environmental footprint of transporting certain products to be consumed far from the place where they are produced. While acknowledging that certain combinations of ingredients can have a beneficial effect on health, biochemically speaking, it should also be noted that no individual ingredient can enter the culinary system without finding a culturally suitable place there. We should remember that the introduction of different food products into our diets over history has been complex, to say the least. As highlighted by

Fournier (2005: 142) in relation to the arrival of American products in the Mediterranean area: "Some (products) were accepted fairly quickly while others had to wait quite some time before finding their place on the table of Mediterranean families." And all following different and specific time frames and ways of adapting in each case. By way of some well-known examples, and talking only of American foods, the pepper was adopted fairly quickly; the tomato and the potato for their part, despite their huge popularity today, faced huge difficulties being accepted when they were introduced; foot-dragging that sometimes went on for centuries.

Therefore while it is easy for someone to consume a certain product for a few days or weeks, even months, as a medical prescription or to carry out a scientific research project, it is much more difficult for that person to integrate the said product into their diet on a permanent basis if it is not part of their cultural food environment or their everyday life. It is here that we talk about "adherence."

So, we could say, after a certain study, that someone has improved their health by adding a handful of walnuts to their daily dietary intake or by starting to use olive oil in their diet. But that is merely eating walnuts or consuming olive oil. It is not the Mediterranean diet. There is a big difference between the two.

Consumption, as part of the Mediterranean diet, cannot be separated from production, distribution, legislation, economics, and other social and cultural factors that have historically been constructed around food in the Mediterranean region. In this regard, the Mediterranean diet is not simply a group of healthy nutrients, but rather a complex web of interdependent cultural aspects that range from nutrition to economics and include law, history, politics, and even religion. This point of view, without a doubt, must be underscored in future debates on food sustainability and the Mediterranean diet and its future prospects and challenges.

In this respect, it should be noted that, like any other food system in its own bio-social context (in which, of course, it has survived, transforming, evolving, and adapting to new needs over centuries), the Mediterranean diet (in any of its regional cultural manifestations) is an unbeatable local resource for societies in the Mediterranean area. However, till now it has not been seen as such or protected (González Turmo & Medina, 2012) by the public bodies of these same societies.

To this end, all this panorama needs to be implemented or improved, for example, within the framework of the Euro-Mediterranean association, to achieve the goal of effective sustainable development in the region as indicated in the report *Mediterranean Strategy for Sustainable Development* (MSSD) issued in 2005 by the UN Environment Program:

> Mediterranean agricultural and rural models, which are at the origins of Mediterranean identity, are under increasing threat from the predominance of imported consumption patterns. This trend is illustrated in particular by the decline of the Mediterranean dietary model despite the recognized positive effects on health. The prospective scenario for the expected impacts of trade liberalization, climate change and the lack of efficient rural policies offers a gloomy picture in some southern and eastern Mediterranean countries, with the prospect of aggravated regional imbalances, deeper ecological degradation and persistent or accrued social instability ... Create a conducive regional environment to help countries develop policies and efficient procedures for the labeling and quality certification of Mediterranean food products and to promote the Mediterranean diet. (UNEP/MAP, 2005).

In this regard, as pointed out by Akkelidou in respect of the Mediterranean area, *the current agricultural and commercial policies are inappropriate (. . .). Therefore it is imperative that these are transformed in order to guarantee the quality, security and availability of food, and the sustainability of production, resources and the environment.*

Nevertheless, the Spanish anthropologist Tomé (2021), reflecting on the recent transformations of the Mediterranean cultural landscapes in Spain, says:

> (. . .) The patrimonialization, popularization, and globalization of a certain conception of (the Mediterranean) diet have turned it into a deterritorialized global phenomenon. As a consequence of this process, it has been necessary to notably increase the production of its ingredients to satisfy its growing demand, which, in turn, has generated "secondary effects" in some Mediterranean environments of Southeastern Spain. If, on the one hand, their wealth has increased and population has been established, on the other hand, the continuity of certain cultural landscapes linked to local knowledge and particular lifestyles has been broken, replacing them with agro-industrial landscapes exclusively at the service of production. This, at the same time, has caused social and environmental inequalities. (Tomé, 2021: 3829).

Following this author again, we have a glossed analysis of the report's data that shows the fast growth of greenhouse fruit and vegetable production in all of the Mediterranean basin: Only Almeria province, in the Spanish Southeast, has almost as many greenhouses as both South and North America. Other Mediterranean countries also have seen a notable greenhouse expansion—both plastic and glass—on their soil: Italy (42,800 hectares),

Turkey (41,300), Morocco (20,000), France (11,500), Israel (11,000), and Egypt (6800). (...) The local avifauna is the first to be affected by the growth of this "plastic sea," as the Almerian West is known, since most of its plants are undercover. Due to this, the whole ecosystem is being affected in one way or the other. If it is true that every landscape continuously shifts, as dynamism is one of its characteristics, (...) a "rupture of consistency" has taken place in the modeling of the landscape. In particular, a specific type of Mediterranean landscape, the product of the use of grass for seminomadic cattle in winter in combination with the production of barley and prickly pear, has been radically replaced with a completely new one, particularly with an agroindustrial model based on a social and environmental imbalance.

In a similar way, we can highlight the text devoted to the plastic sea in Almería, corresponding to the "industrial agriculture" section of the "Le grand mezzé" exhibition (De Laubrie, 2021), which can be visited at the Museum of Mediterranean Cultures and Civilizations (MUCEM) in Marseille, France, between May 2021 and December 2023:

> Since the 1970s, the region of Almeria (Spain) has been transformed, despite its aridity but thanks to its record level of sunshine, into an industrial vegetable garden which supplies many European countries with fruit and vegetables all year round. The pollution is considerable, caused mainly by the use of pesticides and fertilizers, but also by the tons of plastic needed to build the greenhouses. Groundwater is unsuitable for any use. Many workers, most of them immigrants, even illegal immigrants, without legal or physical protection, are employed there in often deplorable conditions.

The words against one of the main productive poles of the Mediterranean region that we can find in the exhibition of the French national museum are certainly harsh and put in parallel the production of Mediterranean agricultural products (whose destination is mainly export) with a strong unsustainability both environmental as well as social.

In this regard, it is apparent that the Mediterranean landscapes that produce the basic goods of the Mediterranean diet are being subjected to a type of agro-industrial pressure that should not be compatible with the diet, the lifestyle, and the landscapes inscribed by UNESCO. At the same time, the study "The challenge of land abandonment after 2020 and options for mitigating measures" by the European Parliament (Andronic et al., 2019) highlights that around 30% (circa 56 million ha) of agricultural areas in the EU are under, at least, a moderate risk of land abandonment. Effective agricultural land abandonment in the EU-27 might total

5 million ha by 2030, or 2.9% of the current EU-27 Utilized Agricultural Area (173 million ha). Inside this panorama, a report from the Union of Farmers and Ranchers Unions of Castilla-La Mancha, Spain (2021) warns of the direct danger of abandonment of 2.3 million agricultural hectares in ten years in Spain, following the prevailing trend to date. The said abandonment is due to, according to the same source, among other causes, the urgent need to balance the food chain so that the work and food and services that farmers and ranchers offer to society are fairly remunerated, and to the need to improve rural infrastructures to reduce the gaps between the rural and urban world.

Conclusion

While it is true that the concept (in full) of the Mediterranean diet has undergone a considerable change in the last two decades, particularly following its recognition by UNESCO as an Intangible Cultural Heritage of Humanity in 2010 (and its corresponding extension in 2013), the different views on the Mediterranean diet, which have been created and developed in different academic disciplines, continue—frequently to cause significant problems for a holistic approach to its epistemological acceptance. This has an impact on the basic definition we are talking about and, consequently, on the interdisciplinary dialog on this matter.

The Mediterranean diet as a concept is (and must be) equivalent to a Mediterranean food system/culinary system. Better knowledge and promotion of this reality as a system would significantly help to ensure greater sustainability and a lower footprint for the production and consumption of *Mediterranean food* in the Mediterranean area, in addition to other possible and known benefits. Local production and consumption seem to have a smaller environmental footprint and serve to support the social, cultural, and economic development of societies and also to ensure greater conservation of the existing heritage and cultural consistency and coherence (from the field to the plate). A coherence that, after the inscription by UNESCO as an intangible heritage, the Mediterranean diet must be protected, not only regarding food consumption, but also regarding production, and, in this framework, take into account the different productive Mediterranean landscapes. Following Tomé (2021) once more, the transformation of the previous traditional landscapes, subordinated to market logics, by breaking with the historical processes that have originated them, produce "Mediterranean" products, but they are no longer "Mediterranean."

The latest definitions of the Mediterranean Diet show it to be an inclusive and interdependent issue. However, we can see how different visions of the Mediterranean diet overlap with one another, without reaching a satisfactory dialog and, on occasion, lead to significant contradictions. These contradictions include, for example, wanting to call the Mediterranean diet "traditional" based on the healthy medical properties of some of its ingredients, or wanting to export it artificially all over the world. From this viewpoint, none of the individual elements that make up this heritage should be considered separately or individually: from production to consumption; from origin to eventual recycling.

We started this chapter with a question in the title related to the potential sustainability, both environmental and social, of the Mediterranean diet. In response to this question, it should be said that as much as we wish to act in favor of environmental sustainability through our diets, this in itself will have no effect if it is not linked to the sociocultural sustainability of the societies in which it must be developed. Environmental sustainability is intrinsically linked to the physical, social, cultural, economic, and political development and well-being of the different populations who act upon it. And nothing can be achieved by acting solely on one isolated aspect of the food chain, be it health, the environment, or any other laudable goal.

From a local Mediterranean perspective, and always as a local consumption model, there is no doubt that the Mediterranean diet thus understood, could be a sustainable resource for the Mediterranean area. For this to happen, however, broader perspectives are needed, along with analyses that are holistic, integrated and interdisciplinary, open and respectful, which enable us to see the eating of food as an everyday, multifaceted occurrence that is equally a biological and a sociocultural act.

References

Aboussaleh, Y., Capone, R., & Bilali, H. (2017). Mediterranean food consumption patterns: Low environmental impacts and significant health—nutrition benefits. *Proceedings of the Nutrition Society, 76*(4), 543—548.

Adinolfi, F., Capone, R., & El Bilali, H. (2015). Assessing diets, food supply chains and food systems sustainability: Towards a common understanding of economic sustainability. In A. Meybeck, S. Redfern, F. Paoletti, & C. Strassner (Eds.), Proceedings of international workshop "Assessing sustainable diets within the sustainability of food systems—Mediterranean Diet, organic food: new challenges. 15–16 Sept 2014, Rome (pp. 167–175). Rome: FAO. Available from http://www.fao.org/3/a-i4806e.pdf#page = 221.

Andronic, C., Derszniak-Noirjean, M., Gaupp-Berghausen, M., Hsiung, C. H., Münch, A., Schuh, B., Dax, T., Machold, I., Schroll, K., & Brkanovic, S. (2019). *Research for AGRI-Committee. The challenge of land abandonment after 2020 and options for mitigating measures*. Strasbourg: European Parliament. Available from https://www.europarl.europa.eu/thinktank/en/document/IPOL_STU(2020)652238.

Berry, E., Dernini, S., Burlingame, B., Meybeck, A., & Conforti, P. (2015). Food security and sustainability: Can one exist without the other? *Public Health Nutrition, 18*(13), 2293–2302.

Boström, M., & Klintman M. (2009). The green political food consumer. Anthropology of food, S5, September 2009. Available from http://aof.revues.org/index6394.html Accessed 15.12.20.

Bottalico, F., Medina, F. X., Capone, R., El Bilali, H., & Debbs. (2016). Erosion of the Mediterranean Diet in Apulia Region, south-eastern Italy: Socio-cultural and economic dynamics. *Journal of Food and Nutrition Research, 4*(4), 258–266. (10.12691).

Burlingame, B., & Dernini, S. (2011). Sustainable diets: The Mediterranean Diet as an example. *Public Health Nutrition, 14*(12A), 2285–2287.

Buttriss, J., & Riley, H. (2013). Sustainable diets: Harnessing the nutrition agenda. *Food Chemistry, 140*, 402–407.

Capone, R., El Bilali, H., Debs, P., Cardone, G., & Berjan, S. (2013). *Nitrogen fertilizers in the Mediterranean Region: Use trends and environmental implications. Fourth international scientific symposium Agrosym 2013, Jahorina, Bosnia and Herzegovina, Faculty of Agriculture* (pp. 1143–1148). University of East Sarajevo.

Crellin F., Harvey J. (2022) Drought hits Mediterranean crops, rest of Europe in good condition. Paris, Reuters, February 21.

Dempsey, N., Bramley, G., Power, S., & Brown, C. (2011). The social dimension of sustainable development: Defining urban social sustainability. *Sustainable Development, 19*, 289–300.

Dernini, S., Berry, E., Serra Majem, L., La Vecchia, C., Capone, R., Medina, F. X., Aranceta, J., Belahsen, R., Burlingame, B., Calabrese, G., Corella, D., Donini, L. M., Meybeck, A., Pekcan, A. G., Piscopo, S., Yngve, A., & Trichopoulou, A. (2016). Med Diet 4.0. The Mediterranean diet with four sustainable benefits. *Public Health Nutrition, 20*(7), 1322–1330.

Eizenberg, E., & Jabareen, Y. (2017). Social sustainability: A new conceptual framework. *Sustainability, 9*(1), 68. Available from https://doi.org/10.3390/su9010068.

Food and Agriculture Organization of the United Nations. (2012a) Greening the economy with agriculture. Working paper 4: utilization. Improving food systems for sustainable diets in a green economy. FAO, Rome. Available from http://www.fao.org/docrep/015/i2745e/i2745e00.pdf.

Food and Agriculture Organization of the United Nations. (2012b). *Towards the future we want. End hunger and make the transition to sustainable agricultural and food systems*. Rome: FAO. Available from http://www.fao.org/docrep/015/an894e/an894e00.pdf.

Food and Agriculture Organization of the United Nations. (2013a). *The state of food and agriculture. Food systems for better nutrition*. Rome: FAO.

Food and Agriculture Organization of the United Nations. (2013b). *Food security and nutrition in the Southern and Eastern Rim of the Mediterranean Basin*. Cairo: FAO.

Food and Agriculture Organization of the United Nations. (2013c). *Tackling climate change through livestock. A global assessment of emissions and mitigation opportunities*. Rome: FAO Source. Available from http://www.fao.org/3/8d293990-ea82-5cc7—83c6—8c6f461627de/i3437e.pdf, Accessed 25 Apr 2017.

Food and Agriculture Organization of the United Nations. (2016) The state of food and agriculture 2016. Climate Change, Agriculture and Food Security. Available at http://www.fao.org/3/a-i6030e.pdf. Accessed 18.05.17.

Fournier, D. (2005). ¿Y si la luz también hubiera venido del oeste para los campesinos del Mediterráneo? In J. Contreras, A. Riera, & F. X. Medina (Eds.), *Sabores del Mediterráneo. Aportaciones para promover un patrimonio alimentario común* (pp. 140–155). Barcelona: IEMed.

Garnett, T. (2013). Food sustainability: Problems, perspectives, and solutions. *Proceedings of the Nutrition Society*, 72, 29–39.

Gómez-Limón, J. A. (2010). Evolución de la sostenibilidad del olivar en Andalucía. Una propuesta metodológica. *Cuides*, 5, 95–140.

González Turmo, I. (1996). Introducción. Modelos ideales y realidad en alimentación. In I. González Turmo, & P. Romero de Solís (Eds.), *Antropología de la alimentación: Nuevos ensayos sobre la dieta mediterránea* (pp. 13–20). Sevilla: Junta de Andalucía/Fundación Machado.

González Turmo, I. (2005). Algunas notas para el análisis de las cocinas mediterráneas. In J. Contreras, A. Riera, & F. X. Medina (Eds.), *Sabores del Mediterráneo. Aportaciones para promover un patrimonio alimentario común* (pp. 44–64). Barcelona: IEMed.

González Turmo I., Medina, F.X. (2012). Défis et responsabilités suite à la déclaration de la diète méditerranéenne comme patrimoine culturel immatériel de l'humanité (Unesco), in Revue d'Ethnoécologie, 2. Available from https://ethnoecologie.revues.org/957 Accessed 06.01.21.

Gussow, J. D. (1995). Mediterranean diets: Are they environmentally responsible? *American Journal of Clinical Nutrition*, 61, 1383S–1389S. Available from https://doi.org/10.1093/ajcn/61.6.1383S.

Hachem F., Capone R., Dernini S., Yannakoulia M., Hwalla H., Kalaitzidis Ch (2016). The Mediterranean Diet: A sustainable consumption pattern. MedTerra 2016. Bari, Ciheam.

Heller, M. C., Keoleian, G. A., & Willett, W. C. (2013). Toward a life cycle-based, diet-level framework for food environmental impact and nutritional quality assessment: a critical review. *Environmental Science & Technology*, 47, 12632–12647.

Hubert, A. (1998). Autour d'un concept: l'alimentation mediterraneenne. *Techniques et Cultures*, 31–32, 153–160.

International Centre for Advanced Mediterranean Agronomic Studies (CIHEAM) and Food and Agriculture Organization of the United Nations (FAO). (2015). *Mediterranean food consumption patterns: diet, environment, society, economy and health. A White Paper Priority 5 of the Expo Milan 2015 Feeding Knowledge Programme*. Rome: FAO.

Lacatsu, C. M., Grigorescu, E. D., Floria, M., Onofriescu, A., & Bogdan-Mircea, M. (2019). The Mediterranean diet: From an environment-driven food culture to an emerging medical prescription. *International Journal of Environmental Research and Public Health*, 16(6), 942. Available from https://www.ncbi.nlm.nih.gov/pmc/articles/PMC6466433/. (accessed: December 13th, 2020).

Laubrie, É. (2021). Industrial agriculture. In Éde Laubrie (Ed.), *Le Grand mezzé (texts of the exhibition)*. Marseille: MUCEM.

Macbeth, H., & Medina, F. X. (2020). New views on the Mediterranean diet. *Nature Food*, 1, 141–142.

Medina, F. X. (2011). Food consumption and civil society: Mediterranean diet as a sustainable resource for the Mediterranean area. *Public Health Nutrition*, 14(12A), 2346–2349.

Medina, F. X. (2015). Assessing sustainable diets in the context of sustainable food systems: socio-cultural dimensions. In A. Meybeck, S. Redfern, F. Paoletti, & P. Strassner

(Eds.), *Assessing sustainable diets within the sustainable food systems. Mediterranean Diet, organic food: new challenges.* Rome: FAO, CREA and International Research Network for Food Quality and Health.

Medina, F. X. (2019a). Sustainable food systems in culturally coherent social contexts: Discussions around culture, sustainability, climate change and the Mediterranean diet. In P. Castro, A. M. Azul, W. Leal Filho, & U. M. Azeiteiro (Eds.), *Climate change-resilient agriculture and agroforestry. Ecosystem services and sustainability* (pp. 189—196). Cham: Springer.

Medina, F. X. (2019b). From healthy to sustainable: Transforming the concept of the Mediterranean diet from health to sustainability through culture. In P. Collinson, I. Young, Ly Antal, & H. Macbeth (Eds.), *Food and sustainability in the twenty-first century: cross-disciplinary perspectives.* Oxford: Berghahn.

Mediterranean Diet (2010). Transnational nomination. Greece, Italy, Morocco, Spain.

Medina, F. X. (2021). Looking for commensality: On culture, health, heritage, and the mediterranean diet. *International Journal of Environmental Research and Public Health, 18,* 2605. Available from https://doi.org/10.3390/ijerph18052605.

Moser, P. (2001). Glorification, disillusionment, or the way into the future? The significance of Local Agenda 21 processes for the needs of local sustainability en. *Local Environment, 6*(4), 453—467. Available from https://doi.org/10.1080/13549830120091734.

O'Kane, G. (2012). What is the real cost of our food? Implications for the environment, society and public health nutrition. *Public Health Nutrition, 15*(02), 268—276.

Rivero, B., Conde, D., & Mariano, L. (2021). Alimentación, sostenibilidad y tradición. Una etnografía de la revitalización gastronómica de Las hurdes (España). In D. Conde, L. Mariano, & F. X. Medina (Eds.), *Gastronomía, cultura y sostenibilidad. Etnografías contemporáneas.* Barcelona: Icaria.

Sáez-Almendros, S., Obrador, B., Bach-Faig, A., & Serra-Majem, L. (2013). Environmental footprints of Mediterranean versus Western dietary patterns: beyond the health benefits of the Mediterranean diet. *Environmental Health: A Global Access Science Source, 12,* 118.

Serra-Majem, L., & Medina, F. X. (2016). The Mediterranean diet as an intangible and sustainable food culture. In V. R. Preedy, & D. R. Watson (Eds.), *The Mediterranean diet: an evidence-based approach* (pp. 37—46). London: Academic Press-Elsevier.

Tilman, T., & Clarck, M. (2014). Global diets link environmental sustainability and human health. *Nature, 515,* 518—522.

Tomé, P. (2021). Unexpected effects on some spanish cultural landscapes of the mediterranean diet. *International Journal of Environmental Research and Public Health, 18.* Available from https://doi.org/10.3390/ijerph18073829.

Trichopoulou, A., Martínez-González, M. A., Tong, T., Forouhi, N. G., Khandelwal, S., Prabhakaran, D., Mozaffarian, D., & de Lorgeril, M. (2014). Definitions and potential health benefits of the Mediterranean diet: Views from experts around the world. *BioMed Central Journal (BMC) Medicine, 12,* 112. Available from https://link.springer.com/content/pdf/10.1186/1741-7015-12-112.pdf.

UNEP/MAP. (2005). Mediterranean strategy for sustainable development: a framework for environmental sustainability and shared prosperity. In: Proceedings of the 14th ordinary meeting of the contracting parties to the convention for the protection of the marine environment and the coastal region of the Mediterranean and its protocols, Athens.

Vanham, D., Del Pozo, S., Pekcan, A. G., Keinan-Boker, L., Trichopoulou, A., & y Gawlika, B. M. (2016). Water consumption related to different diets in Mediterranean cities. *Science of the Total Environment, 573,* 96—105.

CHAPTER FOUR

Sustainability at the menu of Malaysian, Indonesian, and Singaporean top chefs

Jean-Pierre Poulain[1,2], Siti Ramadhaniatun Binti Ismail[3] and Frederic Cerchi[4]

[1]Faculty of Social Sciences, Taylor's University, Kuala Lumpur, Malaysia
[2]Institut Supérieur du Tourisme, de l'Hôtellerie et de l'Alimentation (ISTHIA), Jean Jaurès University, Toulouse, France
[3]School of Food Studies & Gastronomy, Taylor's University, Kuala Lumpur, Malaysia
[4]Taylor's Culinary Institute, Kuala Lumpur, Malaysia

Emergence of awareness

The industrialization and commodification of food first created the conditions for a change in the outlook on local food cultures and created the food heritage movement. It therefore seems to go without saying today that these "food cultures" are in danger and that they must be protected, inventoried, and valued, and in any case designated as heritage... The first effects of industrialization and globalization are multiple and partially contradictory. We see the erosion of traditional food cultures, the introduction of transnational products (pizza, hamburger, French fries, soda, etc.), and the spread of culinary cultures from different regions of the world (Poulain, 2009, 2017). It is in this context that *the nouvelle cuisine* appeared. Its development started in the 1980s in Europe, and its resourcing in the local food cultures created the conditions for the birth of new fine dining restaurants serving locally inspired cuisine. It spread later in Asia, Australia, and South America. The connection with tourism was one of the ways of dissemination. The links between modernization, tourism development, and food cultures have been early studied in different cultural contexts (Bessière, 1998; Bessière et al., 2013; Cohen & Avieli, 2004; Csergo, 1996, 2016; Defer, 1987; Dixit, 2019; Fischler, 1990; Hall, 2003; Medina & Tresserras, 2018; Moulin, 1975; Poulain, 1993, 1997, 2000, 2008; Reynolds, 1993; Scarpato, 2003; Tibère, 1997, 2001).

Food, Gastronomy, Sustainability, and Social and Cultural Development.
DOI: https://doi.org/10.1016/B978-0-323-95993-3.00012-8

© 2023 Elsevier Inc.
All rights reserved.

49

Climate change and environmental concerns have increased the awareness of the social and ecological impacts of the globalized agro-nutritional system and have contributed to the emergence of the notion of local food toward which multiple theoretical perspectives from the human and social sciences converge. If this began in the 1970s with the ecological critique of the industrialization of food production and distribution, it was strengthened and formalized with the development of ethnology and sociology of food, the broadening and thematization of the notion of heritage, and more recently, the theoretical framework of sustainable development (Amilien, 2011; Feenstra, 1997; Lang & Barling, 2012; Poulain, 2021a, 2021b; Robbins, 2015; Sims, 2009).

However, the notions of food heritage, food cultures, and gastronomy have often been used as equivalent, introducing a certain confusion between the different kinds of experience, different categories of actors, and different attitudes toward cuisines. For many authors patrimonialization implicitly transforms a food culture into a gastronomy. Looking at food traditions as a heritage that must be protected is a great transformation of the perspective. Gastronomization refers to another phenomenon that consists of taking a food culture from the bottom of society as a source of inspiration and creativity to produce a *nouvelle cuisine* for fine dining restaurants (Poulain, 2011a, 2011b). Designating it as a heritage leads to doing inventory, protecting, and disseminating. To gastronomize consists of changing the social status of culinary practices.

So, chefs have come to use ingredients, cooking techniques, and consumption methods, which are sources of inspiration. This is called gastronomization. This phenomenon should not be confused with the patrimonialization described above.

Gastronomy presupposes a reversal of the sociological perspective. French gastronomy is an avatar of the fashion system described by Elias (1994), resulting from the process of copying/distancing from the aristocracy and bourgeoisie. In this movement, local foods marked by the need and connection with the ecological niche are the antithesis of gastronomy, which depicts abundance, profusion, and world domination. By putting the notion of culture in the plural, Hoggart (1957) contributes to looking at popular cultures without condescension. In the 1980s, Bourdieu, in La distinction, attempted a sort of synthesis between Elias and Hoggard, reducing the process of distinction to games of distinction between the elites and the rising classes and affirming the autonomy of popular tastes (Bourdieu, 1981). First, the inspiration *nouvelle cuisine* has developed in regional food cultures is the process of relocalization of styles of gastronomy. Second, awareness of the environmental consequences of food choices has led some chefs to take a clear position on the issue of sustainability by transforming their culinary practices. They thus joined reflections that had begun in the academic sector (Allen, 1993). It's a real conceptual revolution in gastronomy that changes the relationship with nature and its role in societies. However, these new positions may have different terms. They can relate to various dimensions: the supply policy, the ratio of products of animal/vegetable origin, the writing of menus, waste management,.... It is therefore the purpose of this chapter to explore how Malaysian, Indonesian, and Singaporean chefs are putting sustainability on their menus.

Socioeconomic contexts

Malaysia is a multicultural society of 35 million people, composed of several ethnic groups (Malay, non-Malay Bumiputera, Chinese, and Indian, plus a few minorities). Each group has its own food culture with its typical dishes and ingredients, dietary taboos and restrictions, dining rituals, form and structure of meals, and symbolic dimensions of food. However, it could appear as a simple question: these racial categories are not totally homogeneous in Malaysia; the Indians may belong to different religions; for instance, they may be Hindus, Muslims, Sikhs, Buddhists, Christians, or members of New Religious Movements; they may speak different mother tongues, for instance, Tamil, Hindi, Urdu, Malayalam; identify more or less strongly with a caste, come from different regions of India, or countries neighboring India, such as Pakistan or Sri Lanka; their families may be in Malaysia for several generations or just arrived. The Chinese may be Buddhists, Taoists, Christians, or Muslim converts; they may speak Hakka, Hookean, Cantonese, Teochew, or Mandarin; they may be Min people, Hakka, Cantonese, or Wu. Furthermore, there are Malaysians in the official "others" category, such as the non-Malay Bumiputera, the Dusun, Iban, and Kadazan. Moreover, the boundaries between these three main races are not totally hermetic; they have a certain porosity, resulting from interpersonal relationships across ethnic boundaries, through friendships, mixed marriages, overlapping religious affiliations, and language competence and usage behind the primary "race" identity, religious conversion, metissage from historical institutionalized mixed marriages (e.g., in the Baba-Nyonya community in historic times), or actual inter-racial breeding with or without conversion (Clammer, 1980; Hirschman, 1987; Tan, 1982), and the rise of individualism within which Malaysians develop personal preferences in choosing from a wide variety of dietary alternatives. In addition, there is also some metissage between the different food cultures. For example, *Nyonya* cuisine from the Malacca region is a combination of Chinese and Malay food cultures, with some influence from the Portuguese. Some restaurants, such as those labeled *Mamak*, which were originally for Tamil Muslims, are now frequented by consumers of all ethnic groups and thus make a solid contribution to the development of a Malaysian mixed-food culture. This means that some dishes and some food practices are commonly shared by or are compatible with more than one ethnic group.

The second characteristic is the high frequency of food consumed outside the home by the urban population in Malaysia, which is one of the highest in the world. Several studies (MOH Malaysia, 2008; Mognard et al., 2021; Poulain et al., 2020) reveal this high incidence of food consumed outside the home and the positive correlation with the level of urbanization. With urbanization, the opportunities for Malaysians to eat out have increased tremendously, and prices are sometimes lower than the cost of a meal made at home. The idea that increased urbanization has resulted in an increase in eating outside the home leads to the assumption that the prevalence of eating out is now of much greater significance within Malaysian food culture and its outcomes for health. We do not claim that eating out could be globally linked with the rise of obesity. Rather, in this research, we assume that there is a typology based on a cluster of practices that make up the ethnic food lifestyles in Malaysia and that some of them relate to obesity (Poulain et al., 2014).

Singapore is also a multicultural society of more than 5 million people, but a large majority, almost 3/4 of the population, is Chinese. The Indian with a share of 9,1%, the Malay with 13,3%, and the Eurasian Metis are the 3 other groups officially recognized[1]. The food supply for food service and restaurants is huge, from street stalls to the international standard Michelin-starred fine dining. The prevalence of eating out is one of the

---

[1] Identity cards in Malaysia like in Singapore use the mention of "race."

highest in the world, with almost one meal out of every two. The high level of economic development and urbanization of the city-state have led to the existence of five-star hotels in the city, and the competition in the hotel and restaurant industry is very high.

Indonesia is an archipelagic state of over 17,000 islands, including Sumatra, Java, Sulawesi, and parts of New Guinea and Borneo. With almost 272 million people, it is the world's fourth-most populous country. According to the sources, the population of Indonesia consists of 1128 to 1340 ethnic groups. But 16 groups represent 85% of the population, and the Javanese themselves make up more than 40%. More than 700 languages are spoken in Indonesia (Ethnologue.com). The modernization is concentrated in the capital Jakarta. At the same time, Bali Island is one of the hotspots of international tourism and is a strong affirmation of its vibrant culture. There are strong cultural links and similarities between these three societies. The multicultural character is common to Singapore and Malaysia, even if the proportion of the communities is not the same. Malaysia and Indonesia share common historical roots, and a majority represent Islam. However, these three countries are very different when it comes to the level of economic development. GDP per capita, for example, in 2021 was USD 4291.8 for Indonesia, USD 11,371.1 for Malaysia, and USD 72,794.0 for Singapore. Fig. 4.1 shows the evolution of GDP since 1960 and how these differences have developed.

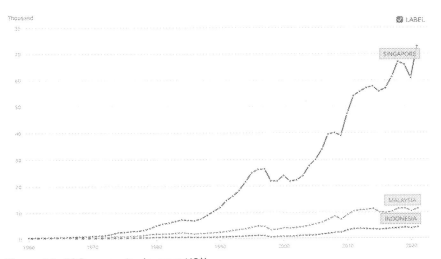

Figure 4.1 GDP per capita (current US$).

Methodology

We have selected 15 restaurants among the top fine dining restaurants in these countries. The criteria of selection were to present a form of interest in sustainability and use it as a commercial argument. From each country, 5 restaurants have been selected, representing 15 establishments altogether. Six of the fifteen establishments selected are or have been present in the Asia 50 Best restaurants. In Malaysia and Indonesia, there is no Michelin guide and therefore no starred restaurants. In 2021 there were 45 starred restaurants for a total of 52 Michelin stars (35 restaurants: 1 star; 7 restaurants: 2 stars, and 3 restaurants: 3 stars), in Singapore. The sample count of 5 Michelin-starred establishments have 2 with 3 stars, 2 with 2 stars, and 1 with 1 star. Location-wise, all Malaysian restaurants are located in Kuala Lumpur; for Indonesian restaurants, 3 are in Jakarta and 2 are in Bali. The national origins of the chefs are quite diverse: for Malaysia, 4 Malaysians and 1 Japanese; for Indonesia, 2 Indonesians, 1 Australian, and 1 Canadian; for Singapore, 1 Singaporean, 2 French, 1 English, and 1 Belgian (see Table 4.1).

The approach consists of using restaurant websites to see how the issue of sustainability fits into the picture. Other works have used the Instagram approach (Mapes & Ross, 2022). We preferred entry via the website to partially erase the difference linked to the level of organization of the restaurant's communication. All establishments do not have the same financial and temporal means to grant it. However, all of them have a website, and even if there are differences in means, they are partly reduced. The levels of observation selected are (Annex 4.1):

- Wording of the carte and menus: The choice of words used in writing the restaurant menu refers explicitly to sustainability, family welfare,..., such as fish coming from small boats, artisanal, ...
- Food sourcing: The sourcing prioritizes food produced in sustainable ways. That includes respect for the environment and efficient use of natural resources; the respect for communities that produce and transform food to ensure that farmers can support themselves and enhance their quality of life. This also includes respect for animals all along the production cycle from the farm or from the sea to the restaurant. Some keywords are organic food, biodynamically grown vegetables, fish sustainability friendly, animal welfare, responsibility. . .

Table 4.1 Sample count of chefs and restaurants.

	Restaurants	Location	Chef	Nationality	Michelin star	Asia's 50 best restaurants
Malaysia	Dewakan	Kuala Lumpur	Darren Teoh	Malaysian	No	#50 (2019—21)
	Akar	Kuala Lumpur	Benny Yeoh	Malaysian	No	No
	Hide	Kuala Lumpur	Ng Yi Shaun	Malaysian	No	No
	Isabel	Kuala Lumpur	Hafizzul Hashim	Malaysian	No	No
	Entier	Kuala Lumpur	Masashi Horiuchi	Japanese	No	No
Indonesia	Mozaic	Bali	Chris Salans	American	No	#1 (2013)
	Merah Putih	Jakarta	Kieran Morland	Australian	No	No
	Blanco	Jakarta	Mandif Warokka	Indonesian	No	No
	Cuca	Bali	Kevin Cherkas	Canadian	No	No
	Locavore	Bali	Ray Adriansyah/Eelke Plasmeijer	Indonesian/ Netherland	No	#21 (2018)
Singapore	Odette	Singapore	Julien Royer	French	3	#8 (2021)
	Les Amis	Singapore	Sebastien Lepinoy	French	3	#23 (2021)
	Labyrinth	Singapore	Han Li Guang	Singaporian	1 (2017—18—19—20)	#40 (2021)
	Jaan	Singapore	Kirk Westaway	English	2	1 (2022)
	St Pierre	Singapore	Emmanuel Stroobant	Belgium	1 (2017—18) 2 (2019—21)	1

Annex 4.1 Definition of sustainability criteria and valorization.

Keywords	Definition	Valorization
Wording of the carte and menu	The word used in writing the menu refers explicitly sustainability, family welfare, fish coming from small boats, artisanal — choice of words used on the restaurant menu in supporting sustainability	5 marks = no occurrence, 10 marks = one occurrence, 15 marks = two occurrences, 20 marks = three and plus occurrences
Food sourcing philosophy	The sourcing prioritizes food produced in a way that respects the environment, makes efficient use of natural resources, ensures that farmers can support themselves, and enhance their quality of life in communities that produce food, including animals as well as people.	Idem
Vegetalization or the greening of menus	The menu used a proportion of dishes based on vegetable products, vegetables, legumes, pulses, roots, grains, etc. Three points are considered; first, the percentage of dishes with plants based on the total of the dishes; Second, the vegetarian versions of some dishes, and Third vegetarian versions (mirroring) of some menus.	Plant based dishes: 5 marks = no specific dish, 10 marks = less than 15%, 15 marks = between 15% and 20%, 20 marks = more than 20%, Vegetarian version of the menu: 5 marks = no vegetarian menu, 15 marks = vegetarian menu, 25 marks = mirroring menus
Respect for seasonality	In Southeast Asia in the region study there is no 4 seasons climate. Seasonality means that products will be used in respect to the natural cycle of production (for example durian/mangosteen/ rambutan - available from Mac-June).	5 marks = no, 10 marks, 15 marks given consensually by the authors

(*Continued*)

Annex 4.1 (Continued)

Keywords	Definition	Valorization
Chef's sustainable values	Values related to sustainability are attached to the chef of the restaurant; and claim in the portrait of the chef as une personal decision choice (location, sourcing, using not expensive products, using traditional techniques, or being creative by using it)	5 marks = no, 10 marks, 15 marks, 20 marks given consensually by the authors
Localization of food	Locally sourced ingredients are items that have been purchased nearby from a farmer, fishmonger, or any other fresh produce creator from within a specific radius (distance) where it will be used or sourced, or from a given geographical area.	5 marks = no, 10 marks, 15 marks, 20 marks given consensually by the authors

- Vegetablization of the menus: The menu offers a proportion of dishes based only on plant products, vegetables, legumes, pulses, roots, grains. Three points are considered: first, the percentage of plant-based dishes in the total dishes; second, the vegetarian versions of some dishes, and third, vegetarian versions (mirroring) of some menus.
- Respect for seasons: In Southeast Asia there is no 4 seasons climate according to the region study. Seasonality means that products will be used in respect for the natural cycle of production (for example, durian, mangosteen, and rambutan are available at their best from March to June).
- Localization of food: Locally sourced ingredients where items are purchased from a farmer, fishmonger, or any other fresh produce creator nearby from within a specific radius (distance) from where it will be used or sourced, or from a given geographical area.

In the second time of the analyse the above criteria have been transformed into quantitative variables (Annex 4.2). This basic quantitative evaluation allows us to do sustainable profiles of restaurants, see definitions and valorization in Table 4.2.

Annex 4.2 Valorization and profiling.

Restaurants	Wording	Sourcing	Vegetablization vegetarian options	Season	Chef's values	Localization
Dewakan	5	5	12,5	10	15	10
Akar	5	5	12,5	15	5	10
Hide	5	5	10	5	5	10
Isabel	5	10	17.5	15	5	10
Entier	5	15	15	10	15	10
Mozaic	5	5	12.5	5	10	10
Merah Putin	5	5	15	5	5	10
Blanco	5	5	12.5	10	15	5
Cuca	10	10	20	10	5	10
Locavore	15	20	17.5	10	20	20
Odette	20	20	20	10	15	10
Les Amis	15	5	7.5	10	15	10
Labyrinth	5	5	7.5	5	15	15
Jaan	5	5	20	10	5	10
St Pierre	5	5	10	10	15	5
Total	115	120	210	140	165	155
Mean	7.67	8.00	14.00	9.33	11.00	10.33
Number of rest. using this criterion	4	5	9	11	10	13

Results

The analysis was first conducted qualitatively, and the position of the restaurant was described. See Table 4.2.

- Wording of the carte and menus: Only four restaurants out of the fifteen make explicit reference to the sustainability theme in their way of writing the menus: two from Indonesia (Cuca, Locavor) and two from Singapore (Odette, Les amis), but none from Malaysia.
- Food sourcing: Four restaurants clearly refer to the philosophy of sourcing using sustainability as a criterion. That is the case of Entier and Odette in Singapore, and Cuca and Locavore in Indonesia, again none in Malaysia. Some restaurants name the farmers and the producers, and present them as partners, sometimes calling them the "best supplier" in the world...
- Vegetalization of the menus: It is the most common shared value, with 14 restaurants using this criterion. However, we can identify different

Table 4.2 Findings.

Restaurants	Wording of carte and menus	Food Sourcing (organic/animal welfare, fish sustainability friendly)	Vegetablization of the menus	Respect for seasons (no season, chef mentioned seasonality)	Chef's values	Localization of food
Dewakan	Not mentioned in the menu	Not mentioned	38% of the menu uses vegetables only **Vegetarian menu** No	"We believe in keeping our Food & Drink Tasting Menus an experience that is approachable for everyone, as we believe this showcase our seasons, such as they are, in a succession of dishes over the evening"	Darren Teoh's expression of cooking with purpose. Each dish is made with ingredients sourced from Malaysia, carefully crafted to encourage deeper appreciation for the land and culture. Darren's philosophy is simple. Be curious, have courage, value your roots.	Each dish is made with ingredients sourced from Malaysia, carefully crafted to encourage deeper appreciation for the land and culture.
Akar	Not mentioned in the menu	Not mentioned	32% of the menu uses vegetables only **Vegetarian menu** No	"Underlined by local produce/seasonality and integrated with modern cooking techniques" "We offer a tasty menu revolving around seasonal quality local produce."	No specification about the Chef	"Underlined by local produce/seasonality and integrated with modern cooking techniques"

(*Continued*)

Table 4.2 (Continued)

Restaurants	Wording of carte and menus	Food Sourcing (organic/animal welfare, fish sustainability friendly)	Vegetablization of the menus	Respect for seasons (no season, chef mentioned seasonality)	Chef's values	Localization of food
Hide	Not mentioned in the menu	Not mentioned	16% of the menu uses vegetables only **Vegetarian menu** No	Not mentioned	No specification about the chef	"To source the best products available in the market for our customers, we work closely with farmers and fishermen to ensure the highest quality."
Isabel	Not mentioned in the menu	"These include fragrant Hom Mali rice from Thailand, handmade Belacan (shrimp paste) from Bintulu and juicy free-range chicken from Bentong, to name just a few."	28% of the menu uses vegetables only **Vegetarian menu** No. But vegetarian options among the dishes	"As such we source only the finest and freshest from local and regional growers and suppliers."	No specification about the Chef	Savor regional cuisine in a sophisticated setting.
Entier	Not mentioned in the menu	"Committed to people, plate and the planet. Our team believes that a good dish is not simply	18% of the menu uses vegetables only **Vegetarian menu,** No. But	"French fare with quality seasonal produce"	"Chef Masashi presents creative French fare with quality seasonal produce, from logical sources near or far. Reflecting the nose-	"From logical sources near or far"

	made up of the best cuts, and the integrity of the ingredients can be protected."	vegetarian options among the dishes			to-tail philosophy, Entier's menu features dishes highlighting specific parts of the animal or vegetable ingredient, or signatures served as a whole."	
Mozaic	Not mentioned in the menu	Not mentioned	5% of the menu uses vegetables only Vegetarian menu offered	Not mentioned	Chef Chris Salans brings his culinary genius to your doorstep to create your own fine dining experience at home with your loved ones. "Known for his culinary creativity and impeccable standards, Chef Chris has developed a selection of dishes respecting most dietary preferences. Prepared as set menus, choose the menu of your liking and let our team of professional chefs and waiters take care of everything ease for you."	My team and I invite you to a unique culinary experience based on local ingredients and flavors with Western techniques of preparation and presentation

(*Continued*)

Table 4.2 (Continued)

Restaurants	Wording of carte and menus	Food Sourcing (organic/animal welfare, fish sustainability friendly)	Vegetablization of the menus	Respect for seasons (no season, chef mentioned seasonality)	Chef's values	Localization of food
Merah Putin	Not mentioned in the menu	Not mentioned	24% of the menu is using vegetables only **Vegetarian menu** No. But vegetarian options among the dishes.	Not mentioned	No specifications about the chef	"Celebrating Indonesia's finest cuisine, culture, craftsmanship, and people, Indonesian classics are served alongside creative dishes that experiment with traditional spices and flavors from across the archipelago."
Blanco	Not mentioned in the menu	Not mentioned	18% of the menu uses vegetables only Vegetarian menu offered	"In our way of nurturing the nature and culture, we explore the Mother Nature to procure the freshest ingredients."	"The exotic of Indonesian flavor and carefully sourced finest seasonal quality local produce are merits of his passion and craftsmanship."	Not mentioned
Cuca	"Hand cut sustainable raw fish, traditional tomato sambal, cucumber scales."	Local, "natural products sourced from across Indonesia — maximizing freshness and	21% of the menu uses vegetables only Vegetarian menu offered	"Highlights of our menu that best reflect the soul of Cuca and are inspired by the	No specification about the chef	"Highlights of our menu that best reflect the soul of Cuca and are inspired by the

Locavore	"A sustainable approach to sourcing ingredients. It's not easy, but makes sense, No imports, no dairy or wheat, gluten free, utilizes everything, less animal protein" "Homemade paprika bitter" "Homemade Brem"	supporting local artisans and farmers." "Locavore's core concept is to celebrate local-only ingredients with social responsibility and sustainable practices. Exclusively working with local ingredients is a conscious decision; it's certainly more challenging, but it's also more rewarding and creative too."	14% of the menu uses vegetables only Vegetarian menu offered	freshest vegetarian local products." Through these ongoing relationships, they ensure their guests the freshest of seasonal produce and ethically fed meat animals. And by working closely with local farmers, Locavore supports sustainability within its community.	Ray and Eelke's ingredient-driven menu celebrates the farmers, fishers, and food artisans of Indonesia.	freshest vegetarian local products." The team is committed to environmental sustainability, both in and out of the kitchen. The iconic sage green paint and recycled wood of all outlets reflect that philosophy. All edible kitchen waste is either fed to pigs or composted for the vegetable garden Launched the Jalan Jalan project, a culinary journey of discovery around Indonesia. Jalan Jalan is a mission to identify local foods and cooking techniques that are in danger of disappearing as food becomes increasingly industrialized.

(Continued)

Table 4.2 (Continued)

Restaurants	Wording of carte and menus	Food Sourcing (organic/animal welfare, fish sustainability friendly)	Vegetablization of the menus	Respect for seasons (no season, chef mentioned seasonality)	Chef's values	Localization of food
Odette	"Organic French Poulette" "Rosemary smoked organic eggs" "Wild Atlantic Turbot"	"Through years of genuine relationships and honest friendships with some of the best boutique producers globally, we have been entrusted to serve the best ingredients in their purest form – right in the heart of a city that is at the crossroads of the world." To our key partners here in Singapore who share with us the passion for bringing only	25% of the menu uses vegetables only Vegetarian menu offered	Odette serves modern French cuisine guided by Chef Julien's lifelong respect for seasonality, terroir, and artisanal producers sourced from boutique producers all over the world.	Guided by a lifelong respect for seasonality, terroir and artisanal produce, Julien has devoted years to forging lasting relationships with some of the world's finest boutique producers, taking pride in offering guests a unique opportunity to taste exceptional ingredients at their peak.	"World's finest boutique producers"

the very best to
the table.
To our farmers
at Pocket
Greens, Kok Fah
Technology
Farm and Farm
delight who
provide us with
biodynamically
grown
vegetables,
microgreens and
herbs from our
island city.
For producing
exceptional
organic Tea Tree
Wild Honey
harvested locally
from conserved
honeybees,
Xavier Tan of
Nutrients. For
his sustainably-
raised pigeons,
Fabien Deneour.
To our
fishermen in

(*Continued*)

Table 4.2 (Continued)

Restaurants	Wording of carte and menus	Food Sourcing (organic/animal welfare, fish sustainability friendly)	Vegetablization of the menus	Respect for seasons (no season, chef mentioned seasonality)	Chef's values	Localization of food
		Hokkaido, Brittany, Scotland and Basque country who epitomize a culture of relentless commitment to incredible produce. To Andori Arregui, Alexandre Navarre, Jacques Cocollos and "One Food Source for sharing their passion for outstanding, sustainably-caught seafood."				

Les Amis	"*Le bar de petite pêche « Ike jime », voilé de champignons, servi en marinière*" "*Les côtes d'agneau de l'Aveyron, arlequins de poivrons doux, jus court.*" "*Le gâteau mollet à l'abricot du Roussillon*"	Not mentioned	7% of the menu uses vegetables only **Vegetarian menu** No	Our ingredients are based on seasonal availability; therefore, our menu and prices are subject to change.	Chef Lepinoy's collection of recipes features a plethora of prized seasonal ingredients sourced from all over France. All his dishes are prepared with finesse and a high level of technicality, displaying true mastery of French gastronomy.	Influenced by pure Parisian haute cuisine, Chef Lepinoy's collection of recipes features a selection of the finest ingredients sourced from France.
Labyrinth	Not mentioned in the menu	Not mentioned	9% of the menu uses vegetables only **Vegetarian menu** no	Not mentioned	Chef LG's culinary philosophy hinges on responsibly understanding produce, ingredients and taste profiles, while championing a locavore cause of supporting local community. Labyrinth's 90% locally sourced menu is a sensorial expression of his childhood food	Labyrinth's 90% loc menu

(*Continued*)

Table 4.2 (Continued)

Restaurants	Wording of carte and menus	Food Sourcing (organic/animal welfare, fish sustainability friendly)	Vegetablization of the menus	Respect for seasons (no season, chef mentioned seasonality)	Chef's values	Localization of food
					memories — especially those of grandma's — a relentless enlivenment of traditional recipes and techniques, and the passion of local farmers.	
Jaan	None	Not mentioned	0% of the menu Vegetarian menu Yes. Mirroring the degustation menu.	An honest reflection of JAAN's philosophy, the result is an innovative cuisine that continues to pay tribute to seasonality and the skills of the world's best gourmet producers.		The tasting menu and their seasonal ingredients have been curated and sourced to provide a distinctive Reinventing British dining experience.
St Pierre	None	Not mentioned	21% of the menu uses vegetables only Vegetarian Meal **Vegetarian menu** no	Menu for reference only: subject to change due to the availability and seasonality of ingredients	Sixteen-year-old Emmanuel Stroobant was well on his way to realizing his childhood dream of	Not mentioned

becoming a lawyer when a stint at one of Belgium's best restaurants sparked off a lifelong passion for the culinary arts. Drawn to the hustle and bustle of kitchen life, Emmanuel subsequently learned the ropes from maestros like Pierre Romayer from three Michelin-starred Maison de Bouche and Francis Dernouchamp from two Michelin-starred l'Hostellerie Saint-Roch.

levels. For some restaurant just 10 to 15 dishes are plant based. Others restaurants propose vegetarian versions of the dishes even more of the menus. Isabelle and Entier in Malaysia; Merah Putih, Cuca, and Locavore in Indonesia; and Odette and Jaan in Singapore are the restaurants that present the most vegetalized menus.

- Respect for seasons is used by 11 restaurants (because the meaning of the concept of different seasons for Asia is a bit blurred): Dewakan, Aka, Isabell, Entier, (4 out of the 5 Malaysian restaurants), Blanco, Cuca, Locavore, Odette, les amis, Jaan, and Saint Pierre.
- Localization of food is the commonly used argument by 13 restaurants. With vegetalization it is possible to identify different levels of intensity. The most important are Dewakan, Aka, Hide, Locavore, and Labyrinth.
- Sustainability expressed as one of the values of the chef is present in the communication of nine restaurants.

One establishment above average for all the criteria is Locavore, which is explicitly positioned on the theme of local and sustainability. It therefore declines all the elements. For the other restaurants the profiles are quite varied; however, it is possible to identify three types using the deviations from the mean.

- The first type plays on the chef-location couple. The chef poses as the spokesperson for the local cuisine. In this case, sustainability is understood as a consequence of the localization of food sourcing and local inspiration.
- The second type emphasizes the vegetalization and greening of menus, and the other components are in a minor mode.
- The third type combines revegetation and supply policy.

One establishment is above average for all the criteria is Locavore, which is explicitly positioned on the theme of local and sustainability. It therefore declines all the elements. If among the other restaurants the profiles are quite varied, it is however possible to identify three types using the deviations from the mean.

- The first type plays on the chef-location couple. The chef poses as the spokesperson for a local cuisine. In this case, sustainability is understood as a consequence of the localization of food sourcing and local inspiration.
- The second type emphasizes the vegetalization and the greening of menus, and the other components are in a minor mode.
- The third type combines revegetation and supply policy.

In restaurants with three Michelin stars, there is a process of personalization, that is to say, highlighting the values of sustainability through the chefs. We also find more often the formulation of dishes with a vocabulary explicitly evoking sustainability.

	Restaurant	Profile	Key dimensions
Malaysia	**Dewakan** Kuala Lumpur		• Localization • Chef'ss values • Vegetablization
	Akar Kuala Lumpur		• Localization • Seasonality
	Hide Kuala Lumpur		• Localization

(*Continued*)

(Continued)

Restaurant	Profile	Key dimensions
Isabel Kuala Lumpur	 ISABEL ━━ Isabel ━━ Sample mean	• Seasonality • Vegetablization
Entier Kuala Lumpur	 ENTIER ━━ Entier ━━ Sample mean	• Sourcing • Chef's values • Vegetablization • Season
Mozaic Bali	 MOZAIC ━━ Mozaic ━━ Sample mean	• Localization • Chef's values • Vegetalization • In minor mode

Indonesia

━━ Indonesia ━━ Sample mean

INDONESIA

(Continued)

Restaurant	Profile	Key dimensions
Merah Putih Jakarta		• Vegetablization • Localization
Blanco Jakarta		• Chef's values • Season • Vegetablization
Cuca Bali		• Vegetablization • Sourcing • Wording • Season

	Restaurant	Profile	Key dimensions
	Locavore Bali		• Sourcing • Localization • Chef's values • Wording • Vegetalization
Singapore 	**Odette** Singapore		• Wording • Sourcing • Vegetalization • Chef's values • Season
	Les Amis Singapore		• Wording • Chef's values • Season

(Continued)

Restaurant	Profile	Key dimensions
Labyrinth Singapore	 LABYRINTH — Labyrinth — Sample mean	• Localization • Chef's values
Jaan Singapore	 JAAN — Jaan — Sample mean	• Vegetablization • Season • Localization
St Pierre Singapore	 ST PIERRE — St Pierre — Sample mean	• Sourcing • Chef's values

Discussion and conclusion

Gastronomy fine dining restaurants can be considered as a reading place of the preoccupations of a society. But because these restaurants receive only a small part of society, an internationalized elite, what they give to see cannot be generalized. However, at the end of this journey through the menus and websites of these 15 restaurants in Southeast Asia, several questions arise:

- *Ambiguities of localization*. Location of products in the sense of geographical proximity and location in the sense of geographical indications of provenance. Both can refer to dimensions of sustainability. The first guarantees a low-carbon impact linked with the transport of products and a contribution to the fallout in the local economy of the value created by gastronomy and tourism. The second refers to the qualities of certain biotopes and the know-how of the human communities that inhabit them, which because of this know-how, produce high-quality food. These two conceptions of localization support local producers in their own way. Either by fixing in a perimeter the economic interactions between actors of the hotel business-catering and agriculture or by delimiting the production in a defined space, they contribute to the extracting of the food system from economic rules, known as "relative advantage." But concerning the environmental impact the transportation is a weakness of the second. Over the past decade, food localization has been connected with issues of food sovereignty (Robbins, 2015). In Singapore, sovereignty is such a concern that the country has launched a vast research program called Singapore 2030, which aims to find the conditions to produce 30% of the food consumed in the city-state by that date. The covid 19 epidemic and the war in Ukraine have reinforced the importance of this issue for all countries in the world. But localism in the sense of consuming foods that are produced within a more or less extended radius comes up against another trend of modernity, which is food cosmopolitanism, particularly developed in Singapore and in the metropolitan areas of Malaysia (KL, Joro Baru, and Penang) or Jakarta in Indonesia.
- *Vegetalization*. The greening of food is a global trend in high-end restaurants. This is clearly seen in most restaurants in the sample and in the three countries. It is an old trend, but it took time to spread. It started in high-end catering with the success of Michel Bras's "Gargouillous of

vegetables," (a 3-Michelin-starred chef, elected as the chef who has the most influenced the stared chefs of the world) at the end of the 1990s (Bras, 1991). In the 1990s, the gastronomic guide "The taste of Scotland" awarded a "Vegetarian welcome" label to fine dinning restaurants that offered vegetarian dishes or menus. With the climate issue, the spread of this trend, the counterpart of which was the reduction in the share of food of animal origin, has accelerated (Drewnowski & Poulain, 2019). This phenomenon is observed in countries where the consumption of animal products is very different. They therefore affect the elites, even in the countries in which the protein transition goes from plant protein to animal protein (Poulain, 2021a, 2021b). Over the past decade, food localization has connected with issues of food sovereignty (Pimbert, 2009; Robbins, 2015). In Singapore, sovereignty is such a concern that the country has launched a vast research program called Singapore 2030, which aims to find the conditions to produce 30% of the food consumed in the city-state by that date. The covid 19 epidemic and the war in Ukraine have reinforced the importance of this issue, for all countries in the world. Vegetarian dishes are also an option for Muslim people in a fine dining restaurant, which can also explain in part the development of vegetablization.

- *The styles of cuisine.* The reference to local food culture is mainly done through products, such as vegetables, herbs, fishes, and spices that are the sources of local flavors. From them comes the Malaysian or Indonesian identities, and the uniqueness of the cuisine. To reveal them, the modern techniques are mobilized, and the dishes sometimes adapted to the dietary requirements. Rare are the ones who like Locavore take the "mission to identify local foods and cooking techniques that are in danger of disappearing." Some academic works have started and contributed to put on heritage the food cultures of these three countries and some connections with the leading chefs are established (Abidin et al., 2020; Duruz & Khoo, 2014; Graezer-Bideau & Kilani, 2012; Omar & Omar, 2018; Ramli et al., 2017; Reddy & van Dam, 2020; Tibere & Aloysius, 2013; Wijaya, 2019).

But in the three countries we find a certain degree of cosmopolitanism, but it is in Singapore that this trend is most important. The revered sources of inspiration are Thai, French, Japanese, Italian... cuisines. Cosmopolitanism that mixes influence in the style of cuisines and in the sourcing is a trend of modernization (Poulain et al., 2020). But for chefs working as ambassadors of gastronomy and promoting products even well produced on the other side of the planet, the question of the carbon footprint will be difficult to solve...

This study on gastronomy and sustainability in Malaysia, Indonesia, and Singapore suggests that given the strong interrelationship in terms of food cultures, languages, and habits of a certain multiculturalism, it would be wise to reactivate an already old theoretical framework, which consists in studying globally (Hefner, 2001; Sparke et al., 2004). This triangular approach, which in our case was not totally intentional, could reveal a complex infrastructure of interrelationships at the level of supplies, specialized labor, and professional training systems.

References

Abidin, M. R. Z., Ishak, F. A. C., Ismail, I. A., & Juhari, N. H. (2020). Modern Malaysian Cuisine: Identity, culture, or modern-day fad? *International Journal of Gastronomy and Food Science, 21*, 100220.

Allen, P. (1993). *Food for the future: Conditions and contradictions of sustainability.* Wiley.

Amilien, V. (2011). From territory to terroir?: The cultural dynamics of local and localized food products in Norway. *Sosiologisk Årbok, 16*(3/4), 85–106.

Bessière, J. (1998). Local development and heritage: Traditional food and cuisine as tourist attractions in rural areas. *Sociologia Ruralis, 38*(1), 21–34.

Bessière, J., Poulain, J. P., & Tibère, L. (2013). L'alimentation au cúur du voyage. Le rôle du tourisme dans la valorisation des patrimoines alimentaires locaux. *Tourisme et recherche*, 71–82.

Bessière, J., Mognard, E., & Tibère, L. (2016). Tourisme et expérience alimentaire. Le cas du Sud-Ouest français. *Téoros. Revue de recherche en tourisme, 2*, 35, 35.

Bourdieu, P. (1984). *Distinction: A social critique of the judgement of taste.* Cambridge, Harvard: Harvard University Press, [1979].

Bras, M. (1991). Le livre de Michel Bras, Rodez, Éditions du Rouergue.

Clammer, J. R. (1980). *Sociology of the Baba Communities of Malaysia and Singapore.* Singapore University Press, Singapore.

Cohen, E., & Avieli, N. (2004). Food in tourism, attraction and impediment. *Annals of Tourism Research, 31*(4), 755–778.

Corbeau, J.-P. (1994). *Goût des sages, sages dégoûts, métissage des goûts. Le métis culturel, International de l'imaginaire* (pp. 164–182). Paris: Babel/Actes Sud, n° 1.

Csergo J. (2016). «Tourisme et gastronomie», Téoros [Online], 35, 2, Online since 12 December 2016, connection on 29 July 2022. URL: http://journals.openedition.org/teoros/2916

Csergo, J. (1996). *L'émergence des cuisines régionales, dans Jean-Louis Flandrin et Massimo Montanari (dir.),* Histoire de l'alimentation (pp. 823–841). Paris: Fayard.

Defer, P. (1987). Tourisme et gastronomie. *The Tourist Review, 42*(3), 7–8.

Dixit, S. K. (2019). *The Routledge handbook of gastronomic tourism.* London: Routledge.

Drewnowski, A., & Poulain, J. P. (2019). What lies behind the transition from plant-based to animal protein? *American Medical Association Journal of Ethics, 20*(10).

Duruz, J., & Khoo, G. C. (2014). *Eating together: Food, space, and identity in Malaysia and Singapore.* Rowman & Littlefield.

Elias, N. (1994). *The civilizing process. Sociogenetic and psychogenetic investigations.* Oxford: Basil Blackwell.

Feenstra, G. W. (1997). Local food systems and sustainable communities. *American Journal of Alternative Agriculture, 12*(1), 28—36.

Fischler, C. (1990). *L'Homnivore*. Paris: O. Jacob.

Graezer-Bideau, F., & Kilani, M. (2012). Multiculturalism, cosmopolitanism, and making heritage in Malaysia: A view from the historic cities of the Straits of Malacca. *International Journal of Heritage Studies, 18*(6), London, Taylors and Francis.

Hall, C. M. (2003). (dir.) *Food tourism around the world: Development, management and markets*. Oxford et Boston: Butterworth-Heinemann.

Hefner, R. W. (Ed.), (2001). *The politics of multiculturalism: Pluralism and citizenship in Malaysia*. Singapore, and Indonesia: University of Hawaii Press.

Hirschman, C. (1987). The meaning and measurement of ethnicity in Malaysia: An analysis of census classification. *J Asian Stud, 46*(3), 555—582.

Hoggart, R. (1957). *The uses of literacy: Aspects of working-class life with special to publications and entertainments*. London, U.K.: Chatto and Windus, Published.

Lang, T., & Barling, D. (2012). Food security and food sustainability: Reformulating the debate. *The Geographical Journal, 178*(4), 313—326.

Mapes, G., & Ross, A. S. (2022). Making privilege palatable: Normative sustainability in chefs' Instagram discourse. *Language in Society, 51*(2), 259—283.

Medina, F. X., & Tresserras, J. (Eds.), (2018). *Food, gastronomy and tourism: Social and cultural perspectives*. Universidad de Guadalajara.

Medina, F. X. (2021). Towards a construction of the Mediterranean diet? The building of a concept between health, sustainability and culture. *Food Ethics, 6*(1), 1—10.

Mognard, E., Tibère, L., Bala, P., & Poulain, J. P. (2021). Migrants as brokers of food heritage making: a case study of the Bario Food and Cultural Festival in Borneo, Sarawak, Malaysia. *Vivência: Revista de Antropologia, 1*(57).

MOH Malaysia (2008). Malaysian Adult Nutrition Survey (2003): Vol 4. Meal Pattern of Adults Aged 18 to 59 Years. Ministry of Health, Putrajaya.

Moulin, L. (1975). *L'Europe à table: Introduction à une psychosociologie des pratiques alimentaires*. Bruxelles: Elsevier Séquoia.

Omar, S. R., & Omar, S. N. (2018). Malaysian heritage food (MHF): A review on its unique food culture, tradition and present lifestyle. *International Journal od Heritage, Art and Multimedia, 1*(3), 1—15.

Pimbert, M. (2009). *Towards food sovereignty* (pp. 1—20). London: International Institute for Environment and Development.

Poulain, J.-P. (1993). Identité régionale et tourisme vert à l'heure de l'Europe. *Tourismes, 2*, 83—98.

Poulain, J.-P. (1997). Le goût du terroir et le tourisme vert à l'heure de l'Europe. *Ethnologie Française*, 18—26, XXVII.

Poulain, J.-P. (2000). Les patrimoines gastronomiques et leurs valorisations touristiques. In R. Amirou, & P. Bachimon (Eds.), *Le tourisme local, une culture de l'exotisme* (pp. 157—183).

Poulain, J.-P., & Neirinck, E. (2004). *Histoire de la cuisine et des cuisiniers*. Lanore.

Poulain, J. P. (2005). *French gastronomy, french gastronomies. Culinary cultures of Europe: Identity, diversity and*. In dialogue, (pp. 157—169). Strasbourg: Council of Europe Pub.

Poulain, J. P. (2008). Gastronomic heritages and their tourist valorisations. *Revue Tourisme October*, 1—18.

Poulain, J. P. (2011a). La gastronomisation des cuisines de terroir: sociologie d'un retournement de perspective. In N. Adell, & Y. Pourcher (Eds.), *Transmettre, quel(s) patrimoine(s)? - Autour du patrimoine culturel immatériel* (pp. 239—248). Michel Houdiard editor.

Poulain, J.-P. (2011b). The sociology of gastronomic decolonisation. In Shanta Nair-Venugopal (Ed.), *The gaze of the West: Framings of the East*. Palgrave Macmillan, 2011.

Poulain, J.-P. (2017). *The sociology of food.* London: Bloomsbury, translation of *Sociologies de l'alimentation.* 2002. Paris: PUF.

Poulain, J.-P. (2012). *Dictionnaire des cultures alimentaires.* Paris: PUF.

Poulain, J. P. (2021a). Food in transition, The place of food in the theories of transition. *The Sociological Review, 69*(3), 702–724.

Poulain, J.-P. (2021b). Biodiversity, ethno-diversity and food cultures: Towards more sustainable food systems and diet. In M. Jacob, & U. P. Albuquerque (Eds.), *Local food plants of Brazil* (pp. 3–18). Springer.

Poulain, J. P., Laporte, C., Tibére, L., Mognard, E., & Ragavan, N. (2020). Malaysian Food Barometer (MFB): A study of the impact of compressed modernization on food habits. *Malaysian Journal of Nutrition.*

Ramli, A. M., Zahari, M. S. M., Halim, N. A., & Aris, M. H. M. (2017). Knowledge on the Malaysian food heritage. *Asian Journal of Quality of Life, 2*(5), 31–42.

Ray, K. (2010). A taste for ethnic difference: American gustatory imagination in a globalizing world. In J. Farrer (Ed.), *Globalization, food and social identities in the Asia Pacific region* (pp. 1–17). Tokyo: Sophia University Institute of Comparative Culture.

Reddy, G., & van Dam, R. M. (2020). Food, culture, and identity in multicultural societies: Insights from Singapore. *Appetite, 149,* 104633.

Reynolds, P. C. (1993). Food and tourism: Towards an understanding of sustainable culture. *Journal of Sustainable Tourism, 1*(1), 48–54.

Robbins, M. J. (2015). Exploring the 'localisation'dimension of food sovereignty. *Third World Quarterly, 36*(3), 449–468.

Scarpato, R. (2003). *Gastronomy as a tourist product: The perspective of gastronomy studies. Tourism and gastronomy* (pp. 65–84). Routledge.

Sims, R. (2009). Food, place and authenticity: Local food and the sustainable tourism experience. *Journal of sustainable tourism, 17*(3), 321–336.

Sparke, M., Sidaway, J. D., Bunnell, T., & Grundy-Warr, C. (2004). Triangulating the borderless world: Geographies of power in the Indonesia–Malaysia–Singapore growth triangle. *Transactions of the Institute of British Geographers, 29*(4), 485–498.

Tan, C. B. (1982). Peranakan Chinese in northeast Kelantan with special reference to Chinese religion. *J Malays Branch R Asiat Soc, 55,* 26–52.

Tibère, L. (1997). Promouvoir le patrimoine gastronomique du Vietnam sur le marché touristique: Contribution à une approche sociologique de la découverte de l'altérité alimentaire. *Études vietnamiennes, 3–4,* 569–598.

Tibère, L. (2001). Valorisation de la culture alimentaire: Le patrimoine gastronomique. In J.-P. Poulain, & M. Teychenné (Eds.), *La Recherche en Tourisme* (pp. 143–146). Paris: Jacques Lanore. (dir.).

Tibere, L., & Aloysius, M. (2013). Malaysia as a food-haven destination: The vision and its sustainability. *Asia-Pacific Journal of Innovation in Hospitality and Tourism, 2*(1), 37–51.

Ueda, H., & Poulain, J. P. (2021). What is gastronomy for the French? An empirical study on the representation and eating model in contemporary France. *International Journal of Gastronomy and Food Science, 25,* 100377.

Wijaya, S. (2019). Indonesian food culture mapping: A starter contribution to promote Indonesian culinary tourism. *J. Ethn. Food, 6,* 9. Available from https://doi.org/10.1186/s42779-019-0009-3.

Further reading

Batat, W. (2020). Pillars of sustainable food experiences in the luxury gastronomy sector: A qualitative exploration of Michelin-starred chefs' motivations. *Journal of Retailing and Consumer Services, 57,* 102255.

Condominas, G. (1980). *L'espace social à propos de l'Asie du Sud-Est.* Paris: Flammarion.

Desjeux, D., & Taponier, S. (1991). *Le sens de l'autre*. Paris: L'Harmattan.

Gourou, P. (1955). Remarques sur les régions écologiques. *Annales Biologiques*, *51*, 125−130.

Petrini, C. (2003). *Slow food: The case for taste*. Columbia University Press.

Rinaldi, C. (2017). Food and gastronomy for sustainable place development: A multidisciplinary analysis of different theoretical approaches. *Sustainability*, *9*(10), 1748.

Tibère, L. (2015). Food as a factor of collective identity: The case of creolisation. *French Cultural Studies*, *26*(4), 1−11.

CHAPTER FIVE

Sustainable diets: intergenerational challenges from the Global South

Mariana Hase Ueta
Technische Universität Dresden, Dresden, Germany

Introduction

The act of eating, beyond merely satisfying nutritional needs, mobilizes transnational food chains, memories, perceptions of social reality, and people's aspirations for a better future. Sustainability adds a complex layer to this universe, making people question the environmental impact of their actions, as food production, transportation, and discards have a global ecological footprint. New perceptions on food emerge from the sustainability discussion, different intergenerational experiences, and expectations for the future. These new images reshape gastronomic practices in the global context.

Nutritional transition is the process of changing diets from traditional patterns (which vary according to cultural context) to diets high in sugar, fat, animal products, and processed foods (Popkin & Shu, 2007). This chapter brings perspectives from the Global South and shows how their development narrative influenced their nutritional transition and changed their consumption patterns. Due to economic development and the expansion of consumption, different generations have had different historical experiences of food availability, access, and consumption. These countries also have to deal with the global environmental impact of the change in their consumption patterns. Food production is considered one of the main polluters among industries, and meat production is pointed to as a high-impact product (World Resource Institute, 2009). At the same time that meat becomes a symbol of an environmentally harmful product, it still carries important social meanings in both contexts. Therefore, meat consumption will be the focus of this discussion.

Meat: the enemy of the environment

The report on Sustainable Diets published by EAT-Lancet Commission (Willett et al., 2019) points out animal products as resource-intensive foods due to the emission of greenhouse gases and water footprint. According to Carvalho (2012), the production of 1 kg of beef generates 44 kg of greenhouse gases and 2400 liters of water. "Food is responsible for 25% of the world's ecological footprint, with the consumption of food of animal origin corresponding to 61% of this value, and 33% refers to meat consumption, thus being the largest contributor" (Carvalho, 2012).

The increase in the consumption of food of animal origin is significant in developing countries since the largest part of the production and consumption comes from these regions. In 2007, Popkin and Shu estimated that by 2020 these countries would be responsible for the production of 63% of meat and 50% of milk while consuming 107 mmt (million metric tons) more meat and 177 mmt more milk than in 1996/98 (Popkin & Shu, 2007). This process can be felt not only through direct consumption but also through the growth of supply and consumption networks, which involve large consumption of resources and a large increase in grain production for animal feed. In Brazil, a big meat producer and exporter country, the agricultural and cattle-raising industries are directly linked to the deforestation of the Amazon since these industries are responsible for 80% of deforestation (Marques, 2019).

Popkin and Shu (2007) identify Brazil and China as central to this debate since they have emerged as major meat consumers. These countries underwent a nutritional transition and increased animal product consumption over the last decades. Both countries occupy places of great meat consumers: China reaching an average of 61 kg per capita per year and Brazil 97 kg per capita per year, as seen in graph 1 (Food and Agriculture Organization of the United Nations FAO, 2016). This increase in the volume consumed has a global environmental impact (Escher, 2019) through the global production and supply chains defined as "teleconnections" (Gill et al., 2015).

China and Brazil's economic development and social transformation in recent decades have impacts that go far beyond their domestic sphere and are branching out around the world, weaving new globalizing networks through their exchanges. According to the Yu et al. (2016) forecast, by 2030, China will need 21% more farmland to meet the growing demand for food generated by population growth, urbanization, increased income, and changing nutritional patterns. Of this new demand, it is estimated that approximately one-third of the supply (34 million hectares) will come from countries, such as Argentina, Brazil, the United States, and Thailand, through grains, cereals, soybeans, and rice.

The production of food of animal origin is associated with the large consumption of resources and emission of greenhouse gases, representing an immense negative environmental impact. The urbanization process and the increase in consumption power bring about changes in consumption and food patterns; these trends point to increasing demand from the global supply networks, which has an environmental impact and export risks (Beck, 1992). From a food supply perspective, the ecological impact can be transferred not only through the purchase of foreign land (land grabbing) but also through its effects, such as pollution and intensive water use (Yu et al., 2016).

The debate about the urgency of the global environmental situation while demanding transnational mitigation efforts often fails to consider the agency of the actors involved in this process. The act of meat consumption becomes morally reprehensible given its negative environmental impact (Nordgren, 2012). However, in the context of developing countries undergoing the nutritional transition, it is imperative to listen to the voices of those at the center of the tension between inclusion in the consumer market and the environmental discourse that condemns the expansion of consumption.

These new consumers have to deal with the negative stigma surrounding meat consumption created by the environmental sustainability discourse, which argues that the processes behind the growth in meat production have a significant impact on the increased emission of greenhouse gases and are linked to deforestation and drought (Gill et al., 2015). Meat consumption also plays a vital role in the Sino-Brazilian trade relationship due to the supply chain where Brazil stands out as a major exporter and China as the largest consumer of agricultural products (Hase-Ueta et al., 2018).

China and Brazil: the taste for meat

In the context of the Global South, it is important to highlight the equation "consumption versus sustainability," the change in behavior resulting from the nutritional transition, and the change in the pattern of food consumption in these countries.

The choice of meat as food is to be analyzed as it occupies a central place in food consumption in China (Watson, 2014) and Brazil (Zaluar, 1982). Despite differences in consumption patterns in these countries, where China shows a preference for pork consumption and Brazil for beef and poultry consumption (Gill et al., 2015), there is convergence as to the important cultural symbolic value. As a marker of class ascension and life improvement, meat is also a crucial element in sociability relations—being part of everyday food—and celebration (for example, in Brazil in barbecue and in China among the main dishes of any banquet).

In the Brazilian context, commensality occupies a central place in establishing, maintaining, and expanding social networks. Meat dishes play an essential role in the celebrations and daily life between friends and family. Food consumption not only establishes social ties but is also a marker of social and class differentiation. While some popular expressions such as *despensa cheia* (full pantry), *mesa farta* (table full of food), and *botar comida dentro de casa* (bring food home) denote the issue of abundance, in contrast, it also evidences the experience of deprivation.

Alba Zaluar's ethnography on family roles in the popular urban classes in Rio de Janeiro in 1980 reports that "food is one of the main vehicles through which the urban poor think of their condition" (Zaluar, 1982, p. 173). They report their experiences from the "abundance" of the beginning of the month when they can "even afford meat," which is a highly valued component of the meal, until the end of the month, when they feel "weak" or in other words, they do not have enough money to buy food and have to deal with hunger.

According to Antonio Candido (1971) in São Paulo's countryside, the *caipira* (countryside person) brought to the attention of the author what he called "psychic hunger." This concept expresses that beyond the nutritional aspect of food, it is important to consider the desires and aspirations related to the agency of choosing and being able to afford to eat what they want. He points out that usually, meat, bread, and milk were considered products that mobilize the aspirational dreams of consumption, which would fulfill not only the nutritional need but also a psychic social longing for an access to them.

In these narratives, it is also evident that beyond the aspirational consumption of certain products, in the imaginary of abundance and prosperity, there is the matter of agency, namely the power to choose what to eat. When asked about their aspirations for class mobility and new consumption patterns, the *caipiras* also came up with the concept of "varied food." This would represent the rich people's way of life, those who do not have to fear hunger and have the purchasing power and access to food as they wish. According to Candido, in the context of the countryside of São Paulo, living at a subsistence level provides the primary conditions for these subjects to survive. However, to "live fully," they still dream of improving their lives and having agency, or in other words, being able to access and have the power to choose what they want to eat.

In China, food consumption patterns followed the country's generational transformations, such as mass-market development. In the last century, China's history has been marked by abrupt changes and overwhelming shifts, from the Cultural Revolution to the Reform and Opening Up periods in which it was possible to see the difference in food availability and consumption. As China has increased its purchasing power and experienced mass-market expansion, it has also undergone a nutritional transition (Popkin & Shu, 2007), resulting in changing food consumption patterns and decreasing numbers of malnourished people in the country between 1999 and 2017 (FAOSTAT, 2015).

The place occupied by food, as well as the change in the meaning of consumption itself, implies the transmission of social values at different historical moments in the country (Jun, 2000). In the ethnography of a village in southern China (Fei, 1938), food consumption was central to the understanding of values and relationships between generations (grandparents, parents, and children) in each family, thus becoming a privileged place for observation among the tensions of values between different generations. According to Guo (2000), the understanding and relationship children develop with food come from their parents and grandparents, who teach them how to recognize the different categories of food, the importance of choosing food properly, and how to differentiate between common and celebratory foods. This process of passing on food knowledge is an important exercise that affirms the authority of older generations over children.

Guo (2000) presents the intergenerational tensions through three different discourses: traditionalists, modernists, and consumerists. Traditionalists are the generation born before the founding of the People's Republic of China in 1949 and are characterized by the centrality of the elements coming from Traditional Chinese Medicine and the search for health through balance. The generation of modernists are the people who were born during the Maoist era. Their discourse was influenced by the increase in disposable family income and exposure to scientific nutritional information. This generation had greater access to education than the previous generation and was impacted by Western medicine and modern nutrition science. Between these two generations, tensions could already be observed: the modernists showed intense concern for healthy meals, food safety, and vitamin-rich diets and disdain for the food choices of the traditionalists because they were not guided by scientific nutritional knowledge.

Added to this equation is the generation of consumerists, who were born in the post-opening period (1978) and during the birth of the mass market. This third generation grew during economic prosperity and the wide availability of products, so their decisions on food consumption were not guided by scarcity, giving greater space for new elements. At the same time, access to products must also be considered in light of the growing role of the mass media—with an emphasis on television—in building the aspirational consumption of these new actors.

The food choices of these consumers are less based on balance and nutrition and more on social values and cultural symbols. The consumer generation, despite being the youngest and less privileged hierarchically within the family, has grown in an environment with a greater diversity of communication channels and consequently has greater access to information about new food products. According to Guo (2000), this generation mobilizes and changes the food-related behavior of previous generations (in China, it is common for young people to live with their parents and grandparents) by presenting them with new products that they do not know about, thus making the analysis of this group indicate consumption trends in the country as a whole.

In the early 1980s the Chinese diet was primarily composed of grains and plants. Food was rationed in urban centers due to the planned economy's impact on the agricultural sector. But the agri-reforms after Mao Zedong's death in 1976 began to show their effects on citizens' diets. Between 1981 and 1987, the consumption of grains and vegetables in cities decreased, and the consumption of cooking oil, meat, poultry, and eggs grew in urban centers between 108% and 182%, and in rural areas, this growth was over 200% (Jun, 2000).

The youth in both countries experienced broader and more diverse access to food consumption compared to the older generation. According to Collins and Hitchings (2012), young people are not only the most susceptible social group to change consumption due to environmental values but also represent the possibility of expanding change.

As the focus of the present chapter is to understand changing behaviors and the tension that meat consumption mobilized in the interviewee's families, I selected students in China and Brazil that became vegetarian or vegan. I conducted semistructured interviews on meat perceptions and how the decision to stop eating meat was understood and discussed in their families, with special attention to the two older generations (parents and grandparents). In these families, the debates around the social value of meat were constant, and it was made clear how different generations perceived meat consumption differently.

I collected data from Shanghai in China and São Paulo in Brazil between 2018 and 2019. The interviews investigated urban food consumption and intergenerational values, focusing on the role of environmental values in changing consumption patterns.

Through these testimonies, I intended to trace the history of the intergenerational life of these families, focusing on issues of social mobility and access to consumption. I believe that the experiences of these individuals offer us perspectives of the agents involved in such profound structural changes in Chinese and Brazilian society, showing the interpretations, meanings, and resignifications of the social practices that I intend to investigate, especially regarding consumption. In this approach, I will contrast and analyze the differences among the three generations concerning consumption and how they interpret the narrative of family prosperity.

Different generations sitting at the same table

In light of the tremendous environmental impact of food consumption and considering the increasing consumption of meat in Brazil and China, it is necessary to research from consumers' perspectives to investigate their perceptions, values, and motivations (Barone et al., 2018).

The gastronomic practices in the Global South are constructed involving both memories of food scarcity and expectations for more sustainable food consumption. In the context of both countries, Brazil and China, meat consumption is the focus of intergenerational discussion. While the older generation brings memories of meat as an aspirational consumption, the younger generation associates meat with its negative environmental impact.

The methodology follows the proposal of Guo (2000), which highlights the need to consider food consumption through the generational key and the family meal as an essential arena of intergenerational discussion. The Chinese and Brazilian economic and political history over the last 100 years (Jun, 2000; Belik, 2019) has been troubled, which has generated different experiences regarding the availability and access to food in each generation (Hase-Ueta, 2019). According to Fei (2015), the dynamism of Chinese culture comes from intergenerational dialog. In other words, the family has a central role in understanding and resignifying the development narratives. Guo concludes that family is a privileged locus to research how different generations have different perspectives and perceptions around diet and food.

In the narratives of the young students (vegetarians, vegans, and other groups related to non-meat consumption), I noticed the constant conflict in formulating and reformulating food consumption. These non-meat consumers were a point of convergence of discussions and focus of constant questioning by their families and networks of friends about the meaning of consuming meat and the value it carries. Through these narratives of the transition of individuals, I investigated the perceptions and values related to meat consumption and the conflicts that emerged intergenerationally in their families and in their network of closer sociability. Moreover, in most of the interviews, the youngsters expressed some criticism concerning the vegan "progress discourse." This discourse often implies that more information about meat consumption and production would lead to a process of awareness directly linked to diet change and the reduction of meat consumption. However, the experiences of the students showed that bringing more information to family and friends did not imply dietary change, so these youngsters were led to think that meat mobilizes other senses besides the unreflected consumption and naturalization of value.

Even though China and Brazil had utterly different historical and socio-economic developments, and culture plays a vital role in the construction of values and perceptions about food, I bring converging narratives to the two case studies. One of the main narratives that emerged was meat consumption as a sign of social mobility across generations. Meat was a scarce product for the generation of grandparents in both China and Brazil. This made the product the target of aspirational consumption, and consequently, fulfilling this desire was a sign that "life was getting better."

The young generation has never been deprived of food choices since their childhood. In the interviews with the students in China, the memories of their grandparents of the Great Famine still resonate in their families' ways of understanding their own food consumption and what is considered a valuable dish. In Brazil, the memories of scarcity also influenced how the older generations communicated what "good food" is. In this way, interviewing vegetarian and vegan students allowed me to analyze the conflicts that emerged when they decided to give up what was seen as an important accomplishment of their families: meat consumption.

Environmental impact emerges among these youngsters' main reasons for choosing not to eat meat. Even though the older generations were usually unwilling to change their food consumption choices and often considered it "young people's thing," the students perceived it as a priority to involve the older generation in the change process toward sustainable development. They also showed respect and understanding when crafting their ways to engage in the topic. Many students reported: "they (older generations) could not afford meat in the past, and thanks to their hard work, the family can have access to it now. How to explain to them that now they should not eat it?". What becomes clear in the narratives is that what is really important for all three generations is their agency as consumers, or in other words, their power to choose what to eat.

Conclusion

The Anthropocene, far from having its impacts attributed to undifferentiated humanity, shows that both responsibility and vulnerability are unequally distributed geographically and in terms of class. In this context, it is imperative to include the perspectives of the Global South, where economic growth and environmental impact are in constant tension. Global South historical experiences guide different ways of understanding and responding to sustainability demands. Brazil, a major exporter of meat and soy for animal feed in China, has to deal with the challenge that environmental sustainability brings to the development of food production chains.

Through the intergenerational narratives around food, it is possible not only to see each generation's different experiences regarding food consumption and availability but also the narratives of social mobility that show the acquired power to choose what to eat.

In the Global South, economic growth, nutrition transition, and sustainability aspirations are entangled in the development and gastronomic narratives. Food products like meat are central to establishing important agro-export partnerships, mobilizing memories of deprivation and class mobility, and dreams of achieving a better life. Sustainability and the role of gastronomy demand a cross-cultural, intergenerational global debate. Rethinking what is valued as "good food" and what represents a "good life" in the context of the Global South is a way of reconstructing a more inclusive narrative of global sustainable development.

References

Barone, Bruna, Nogueira, RosanaMaria, Guimarães, K. átia, & Behrens, Jorge (2018). *Sustainable diet from the urban Brazilian consumer perspective. Food research international.*

Beck, Ulrich (1992). *Risk society: Towards a new modernity.* Munich: Sage Publications.

Belik, Walter (2019). Segurança Alimentar e Nutricional na China: aspectos históricos e atuais desafios. In *Estrangeirização de Terras e Segurança Alimentar e Nutricional: Brasil e China em Perspectiva.* UFPE. Recife, Brazil.

Collins, Rebecca, & Hitchings, Russel (2012). A tale of two teens: disciplinary boundaries and geographical opportunities in youth consumption and sustainability research. In *Area* Vol. 44 No. 2, pp. 193—199.

Candido, Antonio (1971). *Os Parceiros do Rio Bonito: Estudo sobre o caipira paulista e a transformação dos seus meios de vida.* São Paulo: Livraria Duas Cidades.

Carvalho, Aline (2012). Tendência temporal do consumo de carne no Município de São Paulo: estudo de base populacional — ISA Capital 2003/2008. Tese de Mestrado defendida na Faculdade de Saúde Pública da Universidade de São Paulo.

Escher, F.O. (2019). "efeito China" na economia e na agricultura do Brasil. Sul21, 15/01/2018. Disponível em: https://www.sul21.com.br/jornal/o-efeito-china-na-economia-e-na-agricultura-do-brasil/. Last Access: Jan.

Food and Agriculture Organization of the United Nations (FAO). (2016) http://www.fao.org/faostat/en/#compare.

Fei, Xiaotong (1938). *From the soil: The foundations of Chinese Society.* Foreign Language Teaching and Research Press.

Fei, Xiaotong (2015). *Globalization and cultural self-awareness.* Switzerland: Springer.

Gill, Margaret, Feliciano, Diana, Macdiarmid, Jennie, & Smith, Pete (2015). *The environmental impact of nutrition transition in three case study countries, Food Security* (7, pp. 493—504). Springer.

Guo, Yuhua (2000). Family relations: The generation gap at the table. In Jing Jun (Ed.), *Feeding China's little emperors: Food, children, and social change.* Stanford University Press.

Hase-Ueta, Mariana (2019). A Transformação do Consumo e a Mudança nos Padrões Alimentares na China. In *Estrangeirização de Terras e Segurança Alimentar e Nutricional: Brasil e China em Perspectiva*. UFPE. Recife, Brazil.

Hase-Ueta, Mariana, Weins, NiklasWerner, Elias, Lilian de Pellegrini, & Barbieri, Mariana Delgado (2018). Cadeias alimentares globais: Um olhar para as mudanças nos padrões de consumo na China e seus impactos nas relações produtivas no Brasil. In Bicas Brics (Ed.), *Agrarian studies world conference*. Brasil: Brasília.

Jun, Jing (Ed.), (2000). *Feeding China's little emperors: Food, children, and social change*. Stanford University Press.

Marques, Luiz. Abandonar a Carne ou a Esperança. (2019). Disponível em: https://www.unicamp.br/unicamp/ju/artigos/luiz-marques/abandonar-carne-ou-esperanca?fbclid = IwAR2QIAK0FdimxQJCQGRVZb6fdFIFQSSRWpvLE4PQTToxkInCP-V_nk1cmsa4 Acesso em 19 de jul. de 2019.

Nordgren, Anders (2012). *Ethical issues in mitigation of climate change: the option of reduced meat production and consumption. Agriculture environment ethics*.

Popkin, Barry, & Shu, Wen Ng (2007). The nutrition in high — and low — income countries: What are the policy lessons? *The Journal of the International Association of Agricultural Economists, 37*, 199—211, v.

Watson, J. (2014). Meat: A cultural biography in (South) China. In J. Klein, & A. Murcott (Eds.), *Food consumption in global perspective: Essays in the anthropology of food in honour of jack goody*. London: Palgrame Macmillan.

Willett, W., et al. (2019). Food in the anthropocene: The EAT-Lancet Commission on healthy diets from sustainable food systems. *The Lancet, London, 393*(10170), 447—492.

World Resource Institute (2009). https://www.wri.org/resources/data-visualizations/shifting-diets-high-consumers-animal-based-foods-could-significantly.

Yu, Y., et al. (2016). Global implications of China's future food consumption. *Journal of Industrial Ecology, [S. l.], 20*(3), 593—602.

Zaluar, Alba (1982). As mulheres e a direção do consumo doméstico. In S. Koffes, et al. (Eds.), *Colcha de Retalhos*. São Paulo: Brasiliense.

CHAPTER SIX

Exploring tourists and local consumers' attitudes on service automation in restaurant industry: the Spanish fast-food experience

Nela Filimon and Francesc Fusté-Forné
Department of Business, University of Girona, Girona, Spain

Introduction

While automation of the service experience was already growing before the Covid-19 outbreak, the pandemic has accelerated the transition toward contactless hospitality exchanges. As Sigala reports, "developmental trends and adoption of smart destinations and tourism services, AI, robotics and other digital advances are now accelerated to combat the COVID-19 tourism implications" (2020, p. 314). A technology-based offer, illustrated in this chapter by service robots, allows tourists to cocreate experiences in real time (Buhalis & Sinarta, 2019). Hospitality companies are growingly implementing automation (Belanche et al., 2020; Cain et al., 2019) to increase service quality (Choi et al., 2021) and as a strategy toward sustainability, where artificial intelligence is anticipated to transform businesses (Khakurel et al., 2018; Nishant et al., 2020) and contribute to sustainable development (Kar et al., 2022).

The use of service robots in dining experiences, in particular, not only provides a new gastronomic experience but also brings structural changes in terms of employment and customer behavior. According to Fusté-Forné and Jamal, "from being used more in back-room contexts like food preparation in the kitchen in the early stages, service organizations are now implementing robots in frontline service provision" (2021, p. 48). In this fashion, Blöcher and Alt (2021) analyzed, for example, the business processes and subprocesses in the European restaurant industry and identified, for all the stages, potential

---

Food, Gastronomy, Sustainability, and Social and Cultural Development.
DOI: https://doi.org/10.1016/B978-0-323-95993-3.00013-X

© 2023 Elsevier Inc.
All rights reserved.

95

applications of AI and robotics. The authors also found that the greatest number of AI and robotics service solutions corresponded to customer operations and reservations, both as part of the front-of-house business processes, while for the back-of-house processes, the bulk of the service solutions was for food and beverage preparation (p. 537).

This chapter explores the main characteristics and values of service robots in the creation of gastronomic experiences for both tourists and local consumers. In the hospitality industry, service robots have an important contribution in shaping the future role of techno-human tourists (Sigala, 2018, p. 154), as part of ambient intelligence tourism (Buhalis, 2019) in the cocreation of sustainable food tourism experiences. We also present the perspective of the restaurants' customers, with a special focus on their attitudes and beliefs toward the presence of AI and service robots in the restaurant industry, as well as the changes in their habits due to the Covid-19 pandemic. For the empirical exercise, we analyzed the customers of fast-food restaurants in Spain. Data collected with a quota sampling survey, based on gender and age, were analyzed with descriptive and multivariate quantitative techniques. Overall findings add recent evidence on Spain, showing that restaurant customers are also open to AI and service robots. However, generational effects must be taken into account, with younger consumers (18—39 years) being more open to new innovations and technologies and more likely to get greater satisfaction from the interaction with service robots. Furthermore, human functions remain important, a third of the sample defending the presence of humans in attending clients and one third in favor of not substituting any human function by service robots in the restaurants.

The chapters are organized as follows: Section 2 explores the presence of service robots in tourism and the potential for gastronomy experiences; Section 3 discusses recent data and trends on consumers' attitudes toward service robots; Section 4 is about data and method; Section 5 analyzes the data and discusses the results; and Section 6 presents the conclusions.

Service robots in tourism: the potential for gastronomy experiences

The automation of services is one of the key features of postcovid tourism systems, which will consolidate the structural changes that tourism is facing in the 21st century (Buhalis & Sinarta, 2019) and contribute to a sustainable development of tourism (Loureiro & Nascimento, 2021). The automation of services allows businesses to offer real-time experiences and increase both the service experience and the service quality (Ivanov & Webster, 2019a). Previous research anticipates that robots will represent one out of four employees in the tourism industry (Bowen & Morosan, 2018). In particular, tourism systems are, and will be, featured by service robots. As defined by Pransky (1996), "unlike industrial robots, which are typically found in manufacturing environments, service robots (or "serve us" robots, as I like to refer to them) will cater to the masses, millions of end-users in a variety of settings from the hospital to the home, from restaurants to offices" (p. 4). Noone and Coulter (2012) show that by implementing a management system based on robotics technologies, quick-service restaurants (QSRs) can improve their performance. While substantial research focuses on the substitution of the labor force by robots, the authors argue that robotics technologies can enhance employees' cognitive capacity and, hence, QSR's competitive advantage in the market (e.g., through better prediction of demand, improved logistics, and production management, etc.). Apart from preparing better quality food, robotic technology is expected to control cooking and customers' wait time and reduce food waste, an objective of sustainability adopted by many QSRs. The current restaurant landscape shows that the use of service robots in gastronomic experiences is growing around the world (see Seyitoğlu & Ivanov, 2022).

A definition of service robots

According to the International Federation of Robotics (IFR, 2022), a service robot is able to perform autonomous functions without the intervention of humans. Service robots are "programmable, intelligent devices, with a certain degree of autonomy, mobility, and sensory capabilities, designed to perform a certain task" (Ivanov, Webster, & Berezina, 2020, p. 2). This also includes the interactions between robots and humans, and the understanding of robots as social agents (Van Doorn et al., 2017), which do not only conduct repetitive work — for example, fast food preparation (Murphy et al., 2017) — but also complex tasks, such as cleaning and delivery, and also social interaction — undertaken by service robots, which can also communicate with humans (Chi et al., 2020).

In particular, service robots are "adaptable interfaces that interact, communicate, and deliver service to an organization's customers"

(Wirtz et al., 2018, p. 909). The first robotic hotels were implemented in Japan (Reis et al., 2020; Shin & Jeong, 2020), and the adoption of services robots by the hospitality industry includes roles such as hotel concierges or waiters (Webster & Ivanov, 2020). The reasons for a growing robotization of hospitality experiences are, among others, the improvement of transaction times and faster assistance (Belanche et al., 2021), the reduction of costs to businesses and consumers (Webster & Ivanov, 2020), the capacity to provide 24/7 services (Tussyadiah, 2020), and the ability to offer novelty experiences (Zhang & Qi, 2019). In this fashion, Salo and Hakanen (2022) argue that, for example, the "B2B service robots" and their multitask skills are expected to have a positive impact on each of the three dimensions of sustainability: economic (e.g., more productive, less labor costs) (for a comprehensive cost-benefit analysis of robot adoption in hospitality and tourism, see also Ivanov and Webster, 2019c), social (e.g., by taking over routinary and risky tasks, employees can focus more on motivating and satisfactory tasks); and environmental (e.g., powered by renewable energy, etc.).

Automation in food experiences

Previous research (see Tuomi et al., 2021) shows that there are two main roles in the robotization of services: (1) the supportive automation, which supports the service experience; for example, in relation to actions such as credit card payments; and (2) the substitutive automation, which creates a whole robotic experience as it happens with robot bartenders or robot chefs (Fusté-Forné, 2021). Robots can create entertainment experiences. According to Fusté-Forné and Jamal (2021), "roles in service robotics range from automation for novelty, for example, a restaurant that places a personal robot assistant on every table as a way to create a unique experience, to automation for improving service quality, such as in delivering orders efficiently and punctually" (p. 45).

A change in employees' roles and customers' perceptions (Ivanov & Webster, 2019b) is anticipated in the food tourism in future driven by the growing robot automation of service experiences. In this sense, service robots are transforming and will continue to transform tourist experiences, and both academics and practitioners must analyze the customers' emotions derived from the implementation of service robots as social agents (Lu et al., 2019). Service robots can provide competitive advantage and brand positioning (Naumov, 2019; Tung & Law, 2017; Webster & Ivanov, 2020); for example, a Marriott robot concierge (Naumov, 2019) or the Spyce robotic restaurant in Boston (Holley, 2018). Service robots not only enhance customer experiences, but they also improve the service quality perceived by customers (Kabadayi et al., 2019; Mende et al., 2019; Park, 2020; Zhong et al., 2020).

On the opposite end, Tussyadiah (2020) claims that less human contact is more likely to generate dissatisfaction from consumers while enjoying a restaurant experience such as dining, for example (see also Beldona & Kher, 2015; Gursoy et al., 2019). In the same line, Seyitoğlu and Ivanov (2022) have focused on the consumers' dining experience and expectations in 13 robotic restaurants located in 7 countries, encompassing different cultural environments and types of cuisine (India, Thailand, Malaysia, the United States, Nepal, Canada, and Sweden). The analysis of the online reviews posted by the visitors, over a period of 7 years (2012−19), allowed the authors to distinguish between positive aspects (e.g., robots as attractive entertainers for kids, food ordering facilities provided by the usage of tablets, watching the robot cooking the ordered food in some fast-food restaurants, good value for money, etc.), and some shortages, all in all, mainly related to the technological limitations of the robots as they are yet not able to fully replace humans, act with total independence and efficiency, and thus more likely to cause service failures (see pp. 59−63).

Also focusing on diner experiences, Zemke et al. (2020) ran a qualitative study focused on the applicability of robots in quick service restaurant (QSR) operations and found that consumers were somehow in between, pointing both positive and negative aspects about their usage. In the same line, evidence cited by Kelso (2022), based on research data provided by Big Red Rooster, suggests that while 32% of restaurant customers don't want robots to prepare their food, a similar proportion (30%) would be rather skeptical seeing robots preparing their meals or serving them the food on the table, another 38% of the interviewed won't mind actually having robots delivering their order to table, and 41% do support the idea of using the robots for cleaning tasks. All in all, it is nevertheless important to notice that, despite any eventual reticence expressed by some customers, they all coincide in accepting robots as being intrinsically related to the future of the food and restaurant industry.

The morphology of robots

While earlier implementations of robots in hospitality environments paid more attention to the technical aspects of the robots (Tung & Law, 2017), nowadays the focus is on the consumer and tourist experiences (Yeoman, 2012). Here, the morphology of robots is crucial (Chan & Tung, 2019; Fong et al., 2003; McCartney & McCartney, 2020) as a source of cocreation of value in human-robot interactions (Christou et al., 2020; Van Doorn et al., 2017). According to Fusté-Forné and Jamal, "the growing social skills of robots and especially anthropomorphic designs contribute toward enhancing services experiences" (2021, p. 46). In addition, robots are increasingly replacing direct human presence in tourist experiences (Kwok & Koh, 2020).

The form of service robots and the role of anthropomorphism (Kim & McGill, 2011; Qiu et al., 2020; Tung & Au, 2018) are also crucial in securing the acceptance of service robots by employees and customers. In this sense, robots that look like humans will play a growing role in the management and marketing of sustainable robotic service experiences (Murphy et al., 2019). According to Mende et al., "interactions between consumers and humanoid service robots will soon be part of routine marketplace experiences" (2019, p. 535). Also, Fusté-Forné and Jamal state that "a humanoid robot will more likely be trusted to act as a bartender that a patron could interact comfortably with than a more visibly mechanical robot behind the bar counter" (2021, p. 52). In this sense, "as a person becomes more attached to IT emotionally and psychologically, technology gains more influence and control, eventually becoming part of the person's identity, and a new form of techno-human identity" (Sigala, 2018, p. 154). This is also a challenge for future food tourism experiences.

Human-robot interactions in gastronomic experiences

Previous work has paid attention to how robots can complement, or substitute, human employees (McLeay et al., 2021; Tuomi, et al., 2019). The designers of automated gastronomic experiences must carefully consider the employee roles in service automation (Larivière et al., 2017; Li et al., 2019; Marinova et al., 2017; Xu et al., 2020). Robots can support some human tasks and substitute others (Fuentes-Moraleda et al., 2020; Lu et al., 2020; Mingotto et al., 2021) in order that employees can provide a more meaningful guest experience (Naumov, 2019; Rodriguez-Lizundia et al., 2015) based on "tasks that require social skills and emotional intelligence" (Ivanov et al., 2020, p. 505).

In the context of a growing automation in food tourism experiences, businesses are also challenged by the notion of hospitality, and they are required to continue offering hybrid products that facilitate both human–human and human–robot interactions as a basis for the cocreation of hospitality. The automation of service involves the different stakeholders for the good of tourism and the economic, environmental, and sociocultural well-being of the local communities, which can build sustainability (Jamal, 2019). The robotization of service experiences must also consider the role of suppliers in creating technological awareness among customers and tourists.

Service robots and consumer attitudes: recent data and trends

Recent data on world robotics (IFR, 2021) show that the slight slowing trend observed in 2019 in the annual installation of industrial robots worldwide due to the Covid-19 pandemic was almost reversed in 2020 in most of the industries, except for the automotive sector, where, from a pick of 126 thousand units installed in 2018, there was a shrink to 102 thousand units in 2019 and to only 80 thousand units in 2020 (p. 10), and the metal and machinery sector, which went from 44 thousand units in 2018 and 46 thousand units, respectively, in 2019, to 41 thousand installed units in 2020, well below the level registered in 2018 (p. 10). In the food sector, although the number of installed robots in 2020 (12 thousand units) is the same as in 2018 (12 thousand units), data show that the sector has overcome the decreasing trend registered in 2019, with only 11 thousand units of robots installed (IFR, 2021, p. 10).

According to the same source (IFR, 2021), for the service robots data for 2020 show a growing trend of the sales, both for professional and consumer service robots. In this fashion, hospitality (with 15 thousand units), is the fourth sector with the highest number of sales of service robots for professional usage in 2020, after transportation and logistics (44 thousand units), professional cleaning (34 thousand units), and medical robotics (18 thousand units), respectively, while agriculture occupies the fifth position in the ranking, with 7 thousand units of robots sold (p. 39). Automatization of restaurants, which received a special impulse during the pandemic, with the utilization of robots to reduce personal contact, is expected to increase the employment of robots in the coming years, and the same applies to other business sectors' services, such as delivery, cleaning and disinfection, medical support, among others (see IFR, 2021).

Deloitte's (2020) report on the restaurant and food service industry also highlights three main restaurant consumers' trends: ultimate convenience, frictionless digital experiences, and heighted safety in the wake of Covid-19 (p. 3). The first trend, ultimate convenience, is consolidating the increase of food delivery and food takeout, at least once a month, as a dominant pattern of behavior in the postpandemic too, across all generations of consumers surveyed (generations Z, Y, X, boomers and silent); data show that 46% of the consumers interviewed, will not go back to prepandemic food restaurant consumption habits; apart from safety matters, convenience is the most frequent motive (62%), as the consumers value a quick delivery (for 75% of the consumers, the optimal waiting time shouldn't go beyond 30 min), either lower or no delivery fees, and a more convenient location of the pickup points, location which could be other than the restaurant (p. 4).

The second trend refers to consumers' preference for frictionless digital experiences (Deloitte, 2020), with 70% of the surveyed consumers opting for digital interaction when it comes to off-premises delivery; 58% indicated that they would prefer digital ordering from a quick service restaurant, and 57% would do it through a mobile phone delivery app (p. 7). Restaurants and food brands with social media accounts were followed, on average, by 48% of the consumers interviewed, while 40% of them manifested their openness toward driverless or drone delivery options, and 21% of the consumers indicated that they choose a restaurant based on its social media account (p. 7). The third consumer trend, heightened safety in the wake of Covid-19, shows that a consistent signaling of safety measures (e.g., food safety, guests' safety, employees' safety, cleanliness, etc.), also in the postpandemic, is much valued by the customers, contributing to an increase in their trust on restaurants' websites, as indicated by 56% of the frequent diners, compared with 43% of the general customers (p. 12). Moreover, customers also manifested their willingness to pay up to 10% more for safety provisions. In this line, the following are the top 5 safety items customers mostly look for: cleaning of surfaces after each use (87%); personal control of cleaning (85%), application of visible cleaning practices (85%), cleanliness officially certified (84%), and health and safety measures for employees (82%) (see Deloitte, 2020, p. 12). All in all, to face the changes in consumers' food habits and expectations, the survey also revealed several potential changes and innovations to be considered by the restaurant industry (e.g., less dining space, more locations, implementation of express lanes for specific services in stores, increase in driving-through business, use of drones and driverless cars for deliveries in specific areas, customized digital experiences, etc.) to increase speed and convenience.

Haas et al. (2020) analyzed the impact of the Covid-19 pandemic on the US restaurant industry and found that it varied according to the type of restaurant: thus, while pizza chains have either maintained or even increased their sales by up to 5% (in the period between 2019 and 2020), quick service restaurants (QSR) lost between 25%—35% of their sales, and casual dining and fine dining restaurants were the most affected, with loses between 70%—80%, and 75%—85% of their sales, respectively (p. 194). In the same line, restaurants that made it easier for the customers to stay digitally engaged (e.g., online ordering, etc.) were less affected by the pandemic. Furthermore, the study of Haas et al. (2020) presents several actions to be undertaken by the restaurant industry, under two most likely postpandemic scenarios for the US and world economy — virus controlled and virus recurrence — for its recovery and preparation for the so-called next normal (p. 195).

Whatever the postpandemic virus scenario, many of the actions proposed confirm existing evidence (see Deloitte, 2020), such as reassurance and visibility of restaurants' safety protocols, consistent digital presence across platforms, restaurant redesign to increase off-premises dining, drive-through, and pickup lanes, as well as the digitalization of the customer experience, among others (p. 200). Lucas (2021) discusses the case of McDonald's franchise, which, apart from enabling self-service ordering stations since 2020, is also implementing some of the above-mentioned actions by testing the automated drive-through ordering in ten of its Chicago-based restaurants. In relation to the trade-off between human and robotized activities, according to McDonald's CEOs, the use of voice-ordering technologies resulted reliable in about 85% of the cases. There is also common agreement that in the restaurant industry, robots are also expected to help restaurants cope better with the high postpandemic labor costs and the increasing market competition in the sector (Wolff, 2022).

Given the observed trends in restaurant customers, vis-à-vis the presence of the service robots for the execution of given tasks, with some in favor and others against, the present research is also intended to add recent evidence for the Spanish fast-food restaurants. Moreover, we also focus on additional variables, such as gender and age, to give a more comprehensive characterization of the Spanish fast-food consumers' profiles.

Data and method

The data set consists of 645 observations collected in the period between March and April 2022 with a structured questionnaire, from individuals selected based on a quota sampling method, representative of the gender and age structure of the Spanish population (National Institute of Statistics, INE, 2022). Respondents, of age 18 and older, are all Spanish residents. The gender variable is codified in two levels, women (49.8%) and men (50.2%), respectively; age was registered as a continuous variable and, for the purpose of the analysis, was organized in three categories: 18–39 years (34.7%), 40–59 years (33.8%), and 60 and above years of age (31.5%).

A special set of questions was intended to collect information about consumers' preferences of fast-food and their attitudes and beliefs toward the introduction of AI elements in the restaurant industry, such as service robots, automatic order-taking machines, preference for the shape of the robots, among others (see Table 6.1). As it can be seen from the data presented in Table 6.1, half of the sample (51%) stated that they do like fast-food while 49% said no. As for the fast-food preference of their family members, 60% indicated that their family eats fast-food as well, while in 40% of the cases, the answer was no. For the data analysis we combined descriptive and multivariate techniques, and the findings are presented and discussed hereafter and in the discussion section.

Results and discussion

As data in Table 6.1 show, more than half of the sample (58.3%) stated that the robotization of fast-food restaurants is not likely to alter their fast-food consumption behavior. Furthermore, a disclosure by gender of the importance of the automatization process for the consumers' purchasing decision has revealed that most women and men foresee no change in their fast-food purchasing habits (see Fig. 6.1). Age, on the contrary, plays a significant role (Kruskal-Wallis statistics: 48.939, degrees of freedom (d.f.) $= 2$ and P-value $< .001$), with younger

Table 6.1 Consumers attitude toward AI in fast-food restaurants.

Variables	%
Is the automatization of processes in the fast-food restaurants important for your purchasing process?	
• Yes	41.7
• No	58.3
Which human function do you think would be difficult to be replaced by a robot?	
• All functions can be automatized	8.5
• Customers should be attended by humans	29.5
• Humans should attend customers' complaints	8.7
• Humans should be in charge of maintaining the order	11.2
• It depends on the robots' degree of sophistication	9.6
• No human function should be replaced by robots	32.1
• Other	0.5
Is a robot's shape important for you?	
• Yes, I prefer a robot with human shape	15.5
• Yes, I prefer a robot with no human shape	18.9
• indifferent	65.6
Likert-type variables (see also Table 6.2)	
How would you value ordering through an order-taking machine?	
• Totally unsatisfied	22.3
• Somehow unsatisfied	16.1
• Neutral	21.2
• Somehow satisfied	22.5
• Totally satisfied	17.8
How would you value the delivery of your order by an automatic service?	
• Totally unsatisfied	26.5
• Somehow unsatisfied	18.3
• Neutral	26.5
• Somehow satisfied	16.7
• Totally satisfied	11.9
How would you value the elaboration of the dishes by a robot?	
• Totally unsatisfied	37.1
• Somehow unsatisfied	20.6
• Neutral	25.9
• Somehow satisfied	11.2
• Totally satisfied	5.3
Would you go to a fast food restaurant totally robotized?	
• Totally unsatisfied	40.5
• Somehow unsatisfied	18.4
• Neutral	23.4
• Somehow satisfied	9.9
• Totally satisfied	7.8

consumers (18—39 years) being more sensitive to the importance of AI for their purchasing process (57.6%), compared with the older age categories (40—59 years, 41.7%; and 60+ years only 24.1%, respectively). This finding is in line with the data on internet users, which show that in Spain in 2020, for example, the share of internet users by age was distributed as follows: 97.9% (for the age group 20—24 years), 97.3% (25—34 years), 81.5% (55—64 years), and 44.7% (for 64+ years) (Statista, 2022).

As for the shape of the service robots, whether they should look like humans or not, a significant proportion of the respondents (65.6%) declared themselves indifferent to this issue. Statistics also show that there are significant differences between men and women about the robots' shape (Mann-Whitney test = 46,975.5 and P-value = .021): 70.2% of men versus 60.9% of women are indifferent to the shape while, only 15.8% of men prefer robots with no human shape compared with 22.2% of the women; robots with human shape are preferred by 16.9% of the women and only 14.0% of the men. As for the generational differences, statistics point out that there are no significant differences between the three age categories (18—39, 40—59, and 60+ years) about the robots' shape in fast-food restaurants (Kruskal-Wallis test: 3.190 and P-value = .203, 95% confidence interval).

The variable that collected respondents' feedback on the human functions that could be potentially substituted by robots returned rather dispersed opinions, with two main picks: (1) customers should be attended

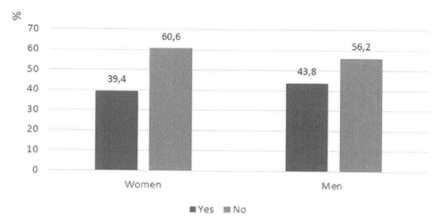

Figure 6.1 Is the automatization process important for your purchasing process?

Service automation in restaurant industry 107

by humans (29.5%), and (2) no human function should be replaced by robots (32.1%). In this case, Mann-Whitney test (49,125.0) and P-value = .292, indicate that there are no significant gender differences. As for the generational patterns, statistics show that there are significant differences between the three age categories: Kruskal-Wallis test: 61.095; d.f.02, P-value < .001 (see Fig. 6.2).

For the remaining questions, although consumers' attitude is overall less concentrated, on average (see Table 6.2), the most impacting issues, on the dissatisfaction side, are related to the elaboration of the food dishes by robots (mean = 2.27) and the decision to go to a totally robotized fast-food restaurant (mean = 2.26). The usage of order-taking machines (mean = 2.97) or of the automatic service delivery (mean = 2.69) seems to have a higher degree of acceptance among the consumers, in both cases

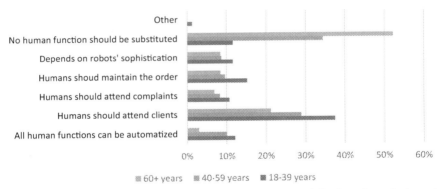

Figure 6.2 Which human function do you think would be difficult to be substituted by robots?

Table 6.2 Main statistics for Likert-type variables.

Variable (1: Totally unsatisfied; 5: Totally satisfied)	Range	Median	Mean	SD
How would you value ordering through an order-taking machine?	1—5	3	2.97	1.41
How would you value the delivery of your order by an automatic service?	1—5	3	2.69	1.34
How would you value the elaboration of the dishes by a robot?	1—5	2	2.27	1.22
Would you go to a fast food restaurant totally robotized?	1—5	2	2.26	1.29

with means above the average, thus being less likely to affect consumers' level of satisfaction with the fast-food restaurants.

For Likert-type variables presented in Tables 6.1 and 6.2, a disclosure by gender shows that, on average, there are no significant differences between women and men, concerning the robotization of the fast-food restaurants. The results of the Mann-Whitney test, for nonnormally distributed data, are summarized in Table 6.3. It can be noticed from the table that, in all cases, a P-value $>.05$ (for a 95% confidence interval) does not allow us to reject the null hypothesis, which assumes that no significant difference exists based on gender.

In this sense, on the opposite end, the analysis shows that customers' attitude toward the applicability of robots in fast-food restaurants depend on age, the generational effect thus having a significant impact on consumers' attitudes and beliefs on this matter as indicated by the statistics for the nonparametric Kruskal-Wallis test (the P-value $<.05$ invites to reject the null hypothesis according to which the three age categories would exhibit a similar behavior) (see Table 6.4). The younger generation (18−39 years) is, on average, more likely to get a higher satisfaction from the interaction with AI elements, such as service robots, while the dissatisfaction of the older consumers (40−59 years, and 60 + years, respectively) is more likely to increase with age (see Fig. 6.3).

Table 6.3 Mann-Whitney test statistics.

Variable	Gender	Mean rank	Mann-Whitney test	Z	P-value (2-tailed)
How would you value ordering through an order-taking machine?	Woman Man	317.19 325.01	50,391.0	-0.491	.623
How would you value the delivery of your order by an automatic service?	Woman Man	316.9 326.04	50,058.5	-0.638	.523
How would you value the elaboration of the dishes by a robot?	Woman Man	312.83 330.12	48,745.0	-1.231	.213
Would you go to a fast food restaurant totally robotized?	Woman Man	308.61 334.31	47,396.5	-1.837	.066

Table 6.4 Kruskal-Wallis test statistics for independent samples.

Variable	Age	Mean rank	Kruskal-Wallis test	d.f.	P-value
How would you value ordering through an order-taking machine?	18–39 40–59 60 +	450.42 313.93 192.14	214.362	2	<.001[a]
How would you value the delivery of your order by an automatic service?	18–39 40–59 60 +	436.57 313.69 207.68	169.961	2	<.001[a]
How would you value the elaboration of the dishes by a robot?	18–39 40–59 60 +	399.94 321.98 239.19	86.024	2	<.001[a]
Would you go to a fast food restaurant totally robotized?	18–39 40–59 60 +	422.04 304.58 233.50	122.908	2	<.001[a]

[a] P-value <.05 significant at 95% confidence interval. Bold values indicate that individuals exhibit a significantly different behaviour, depending on age: the younger age group (18–39 years) is the most open to the interaction with robots or AI, looking forward to this experience and the potential benefits derived from it; older ones (40–59 followed by 60 + years) are, on average, more reluctant, showing that dissatisfaction with the AI is likely to increase with age.

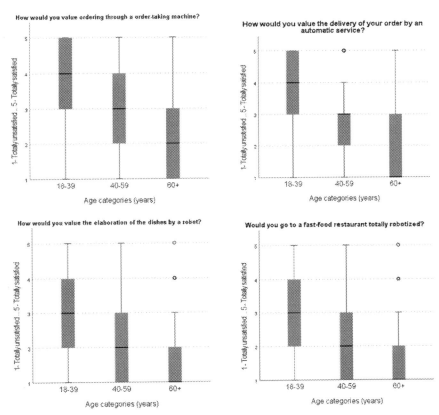

Figure 6.3 Age and satisfaction with service robots.

Conclusions

This chapter shows that service automations based on the implementation of human-robot interactions transform not only the tourist experience (Ivanov et al., 2019; Tussyadiah et al., 2020) but also the customer experience, in general, when it comes to the gastronomic experience. Also, artificial intelligence contributes to the sustainable development of businesses (Kar et al., 2022). A review of the existing literature reveals the following outstanding facts: service robots are expected to play an increasing role in shaping the restaurant industry of the future; customers take it for granted that AI will be an intrinsic part of the gastronomic experience; the use of AI is likely to generate both benefits and shortcomings, and both effects need further analysis for the assessment of their impact both on the restaurants and the customers; the postpandemic customers value most speed and convenience in food service delivery, and to meet these expectations the hospitality sector has to undergo significant restructuring of the business model (e.g., redesign of spaces, alternative locations, pricing policies, etc.); on the customers' side, existing evidence states their willingness to pay an extra fee in exchange for speed and convenience.

Furthermore, AI also challenges the culture of hospitality businesses (Xu et al., 2020) and their sustainable development (Nam et al., 2021). In this sense, "the rising importance of human-robot interactions and activities, however, requires careful attention to collaborative interactions between service actors in the co-creation of services and service encounters, and provision of excellence in service delivery. Robotic environments can disrupt the industry and lead to both positive and adverse behavioral and organizational changes as the traditional framework of a service-provider relationship is being reconstituted" (Fusté-Forné and Jamal, 2021, p. 47). The empirical exercise for Spain, on service robots in the fast-food restaurant industry, is in line with the existing research evidence: on average, more than half of the customers are not going to change their food purchasing habits due to AI; about a third are in favor of preserving the human function when it comes to attending the clients, and another third believes that humans should continue performing all the functions in a restaurant; age, in contrast to gender, has turned out to be a variable with a consistent significant impact on basically all the variables related to AI — the younger generation (18—39 years) is more willing to embrace the presence of AI in the gastronomic experience and is more likely to get a higher level of satisfaction from this experience, compared with the older generations (40−59 and 60 + years). Finally, further research could focus on the effect of AI on other types of restaurants also (e.g., fine dining restaurants) and on the sustainability of the restaurant industry.

References

Belanche, D., Casaló, L. V., & Flavián, C. (2021). Frontline robots in tourism and hospitality: Service enhancement or cost reduction? *Electronic Markets*, *31*(3), 477−492.

Belanche, D., Casaló, L. V., Flavián, C., & Schepers, J. (2020). Service robot implementation: A theoretical framework and research agenda. *The Service Industries Journal*, *40*(3−4), 203−225.

Beldona, S., & Kher, H. V. (2015). The impact of customer sacrifice and attachment styles on perceived hospitality. *Cornell Hospitality Quarterly*, *56*(4), 355−368.

Blöcher, K., & Alt, R. (2021). AI and robotics in the European restaurant sector: Assessing potentials for process innovation in a high-contact service industry. *Electronic Markets*, *31*, 529−551.

Bowen, J., & Morosan, C. (2018). Beware hospitality industry: The robots are coming. *Worldwide Hospitality and Tourism Themes*, *10*, 726−733.

Buhalis, D. (2019). Technology in tourism-from information communication technologies to eTourism and smart tourism towards ambient intelligence tourism: A perspective article. *Tourism Review*, *75*, 267−272.

Buhalis, D., & Sinarta, Y. (2019). Real-time co-creation and nowness service: Lessons from tourism and hospitality. *Journal of Travel and Tourism Marketing*, *36*(5), 563−582.

Cain, L. N., Thomas, J. H., & Alonso, M., Jr (2019). From sci-fi to sci-fact: The state of robotics and AI in the hospitality industry. *Journal of Hospitality and Tourism Technology (Elmsford, N.Y.)*, *10*, 624−650.

Chan, A. P. H., & Tung, V. W. S. (2019). Examining the effects of robotic service on brand experience: The moderating role of hotel segment. *Journal of Travel and Tourism Marketing*, *36*(4), 458−468.

Chi, O. H., Denton, G., & Gursoy, D. (2020). Artificially intelligent device use in service delivery: A systematic review, synthesis, and research agenda. *Journal of Hospitality Marketing and Management*, *29*(7), 757−786.

Choi, Y., Choi, M., Oh, M., & Kim, S. (2021). Service robots in hotels: Understanding the service quality perceptions of human-robot interaction. *Journal of Hospitality Marketing and Management*, *29*(6), 613−635.

Christou, P., Simillidou, A., & Stylianou, M. C. (2020). Tourists' perceptions regarding the use of anthropomorphic robots in tourism and hospitality. *International Journal of Contemporary Hospitality Management*, *32*, 3665−3683.

Deloitte. (2020). The restaurant of the future arrives ahead of schedule. Time to get on board. *Deloitte Development LLC*. Available from https://www2.deloitte.com/content/dam/Deloitte/us/Documents/consumer-business/us-cb-the-restaurant-of-the-future-arrives-ahead-of-schedule.pdf.

Fong, T., Nourbakhsh, I., & Dautenhahn, K. (2003). A survey of socially interactive robots. *Robotics and Autonomous Systems*, *42*(3−4), 143−166.

Fuentes-Moraleda, L., Diaz-Perez, P., Orea-Giner, A., Munoz-Mazon, A., & Villace-Molinero, T. (2020). Interaction between hotel service robots and humans: A hotel-specific Service Robot Acceptance Model (sRAM). *Tourism Management Perspectives*, *36*, 100751.

Fusté-Forné, F. (2021). Robot chefs in gastronomy tourism: What's on the menu? *Tourism Management Perspectives, 37*, 100774.

Fusté-Forné, F., & Jamal, T. (2021). Co-creating new directions for service robots in hospitality and tourism. *Tourism and Hospitality, 2*(1), 43—61.

Gursoy, D., Chi, O. H., Lu, L., & Nunkoo, R. (2019). Consumers acceptance of artificially intelligent (AI) device use in service delivery. *International Journal of Information Management, 49*, 157—169.

Haas, S., Kuehl, E., Moran, J.R., & Venkataraman, K. (2020). How restaurants can thrive in the next normal. In McKensey & Company, *Perspectives on Retail and Consumer Goods*, 8, August 2020 (pp. 194—201). https://www.mckinsey.com/ ~ /media/mckinsey/industries/retail/our%20insights/perspectives%20on%20retail%20and%20consumer%20goods%20number%208/perspectives-on-retail-and-consumer-goods_issue-8.pdf.

Holley, P. (2018). The Boston restaurant where robots have replaced the chefs. *The Washington Post*, 17.

IFR (2021). International Federation of Robotics. *Press Conference World Robotics 2021* (28 October 2021). https://ifr.org/downloads/press2018/2021_10_28_WR_PK_Presentation_long_version.pdf.

IFR. (2022). International Federation of Robotics. *Service Robots*. Available from https://ifr.org/service-robots.

INE. (2022). *National Institute of Statistics*. Available from https://www.ine.es.

Ivanov, S., Gretzel, U., Berezina, K., Sigala, M., & Webster, C. (2019). Progress on robotics in hospitality and tourism: A review of the literature. *Journal of Hospitality and Tourism Technology, 10*, 489—521.

Ivanov, S., Seyitoğlu, F., & Markova, M. (2020). Hotel managers' perceptions towards the use of robots: A mixed-methods approach. *Information Technology and Tourism, 22*(4), 505—535.

Ivanov, S., & Webster, C. (2019a). Conceptual framework of the use of robots, artificial intelligence and service automation in travel, tourism, and hospitality companies. In S. Ivanov, & C. Webster (Eds.), *Robots, artificial intelligence, and service automation in travel, tourism and hospitality* (pp. 7—37). Emerald Publishing Limited.

Ivanov, S., & Webster, C. (2019b). Economic fundamentals of the use of robots, artificial intelligence, and service automation in travel, tourism, and hospitality. In S. Ivanov, & C. Webster (Eds.), *Robots, artificial intelligence, and service automation in travel, tourism and hospitality* (pp. 39—55). Emerald Publishing Limited.

Ivanov, S., & Webster, C. (2019c). Robots in tourism: A research agenda for tourism economics. *Tourism Economics, 26*(7), 1065—1085.

Ivanov, S., Webster, C., & Berezina, K. (2020). Robotics in tourism and hospitality. In Z. Xiang, M. Fuchs, U. Gretzel, & W. Höpken (Eds.), *Handbook of e-Tourism* (pp. 1—27). Springer International Publishing.

Jamal, T. (2019). *Justice and ethics in tourism*. Routledge.

Kabadayi, S., Ali, F., Choi, H., Joosten, H., & Lu, C. (2019). Smart service experience in hospitality and tourism services: A conceptualization and future research agenda. *Journal of Service Management, 30*, 326—348.

Kar, A. K., Choudhary, S. K., & Singh, V. K. (2022). How can artificial intelligence impact sustainability: A systematic literature review. *Journal of Cleaner Production*, 134120.

Kelso, A. (2022). Study: One-third of diners don't want to see robots preparing their food. *Restaurant Drive*, April 11. Available from https://www.restaurantdive.com/news/one-third-of-diners-dont-want-to-see-robots-prepare-their-food/621721.

Khakurel, J., Penzenstadler, B., Porras, J., Knutas, A., & Zhang, W. (2018). The rise of artificial intelligence under the lens of sustainability. *Technologies, 6*(4), 100.

Kim, S., & McGill, A. L. (2011). Gaming with Mr. Slot or gaming the slot machine? Power, anthropomorphism, and risk perception. *Journal of Consumer Research, 38*(1), 94−107.

Kwok, A. O., & Koh, S. G. (2020). COVID-19 and Extended Reality (XR). *Current Issues in Tourism*, 1−6.

Larivière, B., Bowen, D., Andreassen, T. W., Kunz, W., Sirianni, N. J., Voss, C., Wünderlich, N. V., & De Keyser, A. (2017). Service Encounter 2.0": An investigation into the roles of technology, employees and customers. *Journal of Business Research, 79*, 238−246.

Li, J. J., Bonn, M. A., & Ye, B. H. (2019). Hotel employee's artificial intelligence and robotics awareness and its impact on turnover intention: The moderating roles of perceived organizational support and competitive psychological climate. *Tourism Management, 73*, 172−181.

Loureiro, S. M. C., & Nascimento, J. (2021). Shaping a view on the influence of technologies on sustainable tourism. *Sustainability, 13*(22), 12691.

Lu, L., Cai, R., & Gursoy, D. (2019). Developing and validating a service robot integration willingness scale. *International Journal of Hospitality Management, 80*, 36−51.

Lu, V. N., Wirtz, J., Kunz, W. H., Paluch, S., Gruber, T., Martins, A., & Patterson, P. G. (2020). Service robots, customers and service employees: What can we learn from the academic literature and where are the gaps? *Journal of Service Theory and Practice, 30*(3), 361−391.

Lucas, A. (2021). McDonald's is testing automated drive-thru ordering at 10 Chicago restaurants. *CNBC*, June 2. Available from https://www.cnbc.com/2021/06/02/mcdonalds-tests-automated-drive-thru-ordering-at-10-chicago-restaurants.html.

Marinova, D., de Ruyter, K., Huang, M. H., Meuter, M. L., & Challagalla, G. (2017). Getting smart: Learning from technology-empowered frontline interactions. *Journal of Service Research, 20*(1), 29−42.

McCartney, G., & McCartney, A. (2020). Rise of the machines: Towards a conceptual service-robot research framework for the hospitality and tourism industry. *International Journal of Contemporary Hospitality Management, 13*, 3835−3851.

McLeay, F., Osburg, V. S., Yoganathan, V., & Patterson, A. (2021). Replaced by a robot: Service implications in the age of the machine. *Journal of Service Research, 24*(1), 104−121.

Mende, M., Scott, M. L., van Doorn, J., Grewal, D., & Shanks, I. (2019). Service robots rising: How humanoid robots influence service experiences and elicit compensatory consumer responses. *Journal of Marketing Research, 56*(4), 535−556.

Mingotto, E., Montaguti, F., & Tamma, M. (2021). Challenges in re-designing operations and jobs to embody AI and robotics in services. Findings from a case in the hospitality industry. *Electronic Markets, 31*(3), 493−510.

Murphy, J., Gretzel, U., & Pesonen, J. (2019). Marketing robot services in hospitality and tourism: The role of anthropomorphism. *Journal of Travel & Tourism Marketing, 36*(7), 784−795.

Murphy, J., Hofacker, C., & Gretzel, U. (2017). Dawning of the age of robots in hospitality and tourism: Challenges for teaching and research. *European Journal of Tourism Research, 15*, 104−111.

Nam, K., Dutt, C. S., Chathoth, P., Daghfous, A., & Khan, M. S. (2021). The adoption of artificial intelligence and robotics in the hotel industry: Prospects and challenges. *Electronic Markets, 31*(3), 553−574.

Naumov, N. (2019). The impact of robots, artificial intelligence, and service automation on service quality and service experience in hospitality. In S. Ivanov, & C. Webster (Eds.), *Robots, artificial intelligence, and service automation in travel, tourism and hospitality* (pp. 123−133). Emerald Publishing Limited.

Nishant, R., Kennedy, M., & Corbett, J. (2020). Artificial intelligence for sustainability: Challenges, opportunities, and a research agenda. *International Journal of Information Management, 53*, 102104.

Noone, B. M., & Coulter, R. C. (2012). Applying modern robotics technologies to demand prediction and production management in the quick-service restaurant sector. *Cornell Hospitality Quarterly, 53*(2), 122−133. Available from https://doi.org/10.1177/1938965511434112.

Park, S. (2020). Multifaceted trust in tourism service robots. *Annals of Tourism Research, 81*, 102888.

Pransky, J. (1996). Service robots—how we should define them? *Service Robot: An International Journal, 2*(1), 4−5.

Qiu, H., Li, M., Shu, B., & Bai, B. (2020). Enhancing hospitality experience with service robots: The mediating role of rapport building. *Journal of Hospitality Marketing & Management, 29*(3), 247−268.

Reis, J., Melão, N., Salvadorinho, J., Soares, B., & Rosete, A. (2020). Service robots in the hospitality industry: The case of Henn-na hotel, Japan. *Technology in Society, 63*, 101423.

Rodriguez-Lizundia, E., Marcos, S., Zalama, E., Gómez-García-Bermejo, J., & Gordaliza, A. (2015). A bellboy robot: Study of the effects of robot behaviour on user engagement and comfort. *International Journal of Human-Computer Studies, 82*, 83−95.

Salo, M., & Hakanen, T. (2022). How multi-purpose B2B service robots enhance sustainability? *MURO*. https://www.murorobotics.fi/post/how-multi-purpose-b2b-service-robots-enhance-sustainability.

Seyitoğlu, F., & Ivanov, S. (2022). Understanding the robotic restaurant experience: A multiple case study. *Journal of Tourism Futures, 8*(1), 55−72.

Shin, H. H., & Jeong, M. (2020). Guests' perceptions of robot concierge and their adoption intentions. *International Journal of Contemporary Hospitality Management, 32*, 2613−2633.

Sigala, M. (2018). New technologies in tourism: From multi-disciplinary to anti-disciplinary advances and trajectories. *Tourism management perspectives, 25*, 151−155.

Statista (2022). *Percentage of internet users in Spain in 2020, by age*. https://www.statista.com/statistics/751672/share-of-internet-users-in-spain-by-age.

Tung, V. W. S., & Au, N. (2018). Exploring customer experiences with robotics in hospitality. *International Journal of Contemporary Hospitality Management, 30*, 2680−2697.

Tung, V. W. S., & Law, R. (2017). The potential for tourism and hospitality experience research in human-robot interactions. *International Journal of Contemporary Hospitality Management, 29*(10), 2498−2513.

Tuomi, A., Tussyadiah, I., & Stienmetz, J. (2019). Leveraging LEGO Serious Play to embrace AI and robots in tourism. *Annals of Tourism Research, 81*, 102736.

Tuomi, A., Tussyadiah, I. P., & Stienmetz, J. (2021). Applications and implications of service robots in hospitality. *Cornell Hospitality Quarterly, 62*(2), 232−247.

Tussyadiah, I. (2020). A review of research into automation in tourism: Launching the Annals of Tourism Research Curated Collection on Artificial Intelligence and Robotics in Tourism. *Annals of Tourism Research, 81*, 102883.

Tussyadiah, I. P., Zach, F. J., & Wang, J. (2020). Do travelers trust intelligent service robots? *Annals of Tourism Research, 81*, 102886.

Van Doorn, J., Mende, M., Noble, S. M., Hulland, J., Ostrom, A. L., Grewal, D., & Petersen, J. A. (2017). Domo arigato Mr. Roboto: Emergence of automated social presence in organizational frontlines and customers' service experiences. *Journal of Service Research, 20*(1), 43−58.

Webster, C., & Ivanov, S. (2020). *Robots in travel, tourism and hospitality: Key findings from a global study*. Zangador.

Wirtz, J., Patterson, P. G., Kunz, W. H., Gruber, T., Lu, V. N., Paluch, S., & Martins, A. (2018). Brave new world: Service robots in the frontline. *Journal of Service Management*, *29*, 907–931.

Wolff, R. (2022). Restaurants turn to robots and automation as labor shortages continue. *eMarketer*, April 13. https://www.emarketer.com/content/restaurants-turn-robots-automation-labor-shortages-continue.

Xu, S., Stienmetz, J., & Ashton, M. (2020). How will service robots redefine leadership in hotel management? A Delphi approach. *International Journal of Contemporary Hospitality Management*, *32*, 2217–2237.

Yeoman, I. (2012). *2050-tomorrow's tourism*. Channel View Publications.

Zemke, D. M. V., Tang, J., Raab, C., & Kim, J. (2020). How to build a better Robot. . . for quick-service restaurants. *Journal of Hospitality and Tourism Research*, *44*(8), 1235–1269.

Zhang, Y., & Qi, S. (2019). User experience study: The service expectation of hotel guests to the utilization of ai-based service robot in full-service hotels. In F. H. Nah, & K. Siau (Eds.), *International conference on human-computer interaction* (pp. 350–366). Springer.

Zhong, L., Sun, S., Law, R., & Zhang, X. (2020). Impact of robot hotel service on consumers' purchase intention: A control experiment. *Asia Pacific Journal of Tourism Research*, *25*(7), 780–798.

CHAPTER SEVEN

Sustainable restaurants in Barcelona (Spain): identity and sustainability in local cuisine

Manuela Alvarenga Nascimento
Department of Social Anthropology, Universidade Federal de Goiás, Goiânia, Brazil

Introduction

Food sustainability is a growing concern globally. However, it is an issue where it is not always possible to establish a clear link between the food production and consumption chain and what is prepared in the kitchen. Variables such as types of ingredients or knowledge about food preparation are other key aspects to consider when analyzing sustainability. Ingredient awareness, sensorial and nutritional properties, the path to the plate, and how they are prepared are key aspects in the implementation of sustainable kitchens.

We understand that food sustainability is not a concept that has been fully and permanently appropriated anywhere in the world. It is a rather complex process that is taking place in different parts of the world, with specificities in each location. However, it is an interdependent phenomenon since environmental conditions connect territories, even when these are politically separated. This process involves a number of social stakeholders that are under constant tension and dispute, which creates a dynamic of repeated advances and setbacks.

When it comes to sustainability processes, direct engagement from institutions such as the state, local governments, and companies is crucial. Furthermore, it is important to deepen the understanding of citizens' individual responsibilities. The joint efforts of society as a whole and individuals in their private lives are central to achieving food sustainability. A restaurant's day-to-day work can also reflect attitudes and concerns regarding sustainability. Chefs who strive for actions that do not harm the environment and seek information and acquire knowledge on environmental balance will be important actors in achieving sustainability.

Food sustainability, as understood in the context of a restaurant, is expressed in a long list of issues to be taken into account. The physical and architectural structure of the venue, where natural light and air circulation are used effectively, energy-efficient equipment, and "eco-friendly" cleaning supplies are just a few examples. The reduction or elimination of plastics is a particularly important point, as well as garbage or waste management, in addition to a comprehensive use of ingredients to help avoid waste. This is not an exhaustive list of restaurant obligations to achieve sustainability, but they are essential to this process. Moreover, there is the challenge of obtaining all these services at affordable prices while ensuring reliability in the supply chain.

In addition to this material structure, the key element in this type of activity—the restaurant industry—is food itself. Food is composed of both material and immaterial elements. Material elements include how crops are grown, how the food is preserved, its organoleptic properties, etc., while the immaterial aspects are the meanings, knowledge, and cultural traits that shape the defining material properties. We understand that each food will have its own physical properties of flavor, texture, and nutrients as a result of factors such as biological species, farming techniques, chemical use, preservation processes, mode of transportation, and storage conditions. All these aspects are crucial in the final result of the food products, and each depends on knowledge, values, and worldviews that determine the choice of species, varieties, techniques, and sales that may or may not be sustainable.

Hence, we understand that the local actors' perception of sustainability is essential to implementing sustainable food processes. For that reason, sustainable food is the material expression of innumerable sustainable social relationships that have constituted the food in a specific way and not in any other. In this context, it should be considered that the perception of sustainability may vary according to the sociocultural context in which each food is introduced.

We agree with Oosterver et al. (2010: 27), who define sustainable food as having: natural characteristics, or one that has not been adulterated and has been produced using natural processes; safety characteristics, such as not containing traces of pesticides and other chemicals; comes from farms that take into account animal welfare; and is the result of processes that do not harm the environment. In addition, the slow food movement's proposal for "good, clean, and fair" food refers to locally grown food produced with environmentally friendly techniques, having nutritional and organoleptic characteristics that are pleasant and native, and produced in conditions that ensure social justice to producers, complementing the definition.

Accordingly, a sustainable restaurant can be understood as one that has a physical and architectural structure that maximizes the use of available natural resources, has a sustainable food supply chain, minimizes the use of plastic, and reduces and properly manages food waste. Additionally, the specific social and cultural contexts in which sustainable restaurants exist should be considered, as they have their own perception of sustainability, in line with their culture, values, traditions, and environment, as well as their own way of experiencing the process.

The last aspect, experience, is the focus of this research. The perception and local experience of sustainability make this concept rich and diverse. It allows each location to express its concerns according to its social, cultural, and environmental reality and, building upon this, find possible solutions focused on ensuring decent living conditions and abundant biodiversity for future generations. The fact that sustainability, as an international phenomenon, does not have a globally accepted definition could be explained by its cultural element. From this perspective, not reaching a consensus is normal, given the cultural richness and variety found in different societies around the world, which are expressed through different notions of sustainability. Regardless, the focus remains on a balanced use of natural resources and their preservation for future generations.

The issues that underlie sustainable restaurants comprise a broad and intense debate in the social sciences, especially food production and consumption (Goodman, 1999; Guivant, 2002; Murdoch & Miele, 1999; Watts & Goodman, 1997). Mainstream food production and consumption practices are influenced by industries. It is challenging to consider all the characteristics mentioned above for sustainable foods in different social contexts, but they seek to do so.

According to Sagarin et al. (2012), some food transition processes are a work in progress. They require structural changes and lead to the emergence of new forms of production and consumption. These processes are carried out by actors interested in changing the established rules that regulate production, sales, and consumption. According to the authors, these processes reorganize the dynamics of interaction among key actors, such as government authorities, producers, industries, retailers, and consumers. They also generate changes in opinion and behavior, as well as a redefinition of societal values, policies, and practices. Consequently, practices arise, introducing new values and guidelines, such as quality and safety, environmental protection, and animal care, among others, which transcend rationalization and efficiency criteria.

Sustainable restaurants are among the actors seeking structural changes and new forms of production and consumption. Its course of action is guided by ideas and values that transcend efficiency and capitalist profit. In addition to the structural changes, they seek to include concerns for the environment, social justice, and food quality. For this reason, they also identify themselves with what can be understood as a form of "political entrepreneurship."

> Political entrepreneurship, [...] encourages enterprise while advocating for a politicization of food practices. In other words, it advocates for business practices permeated by ideas and values that surpass the basic rationale of accumulation and competition. Political entrepreneurship is set in motion by individuals with critical and creative skills who oppose the political and economic establishment of mainstream food practices, fundamentally influenced by the food industry. (Nascimento, 2014: 127)

Therefore, these enterprises are characterized as innovative businesses that operate in a scenario dominated by a capitalist industrial market and, therefore, under circumstances that are not always favorable and impose many challenges.

Sustainable restaurant: the chef as a key social actor

In a sustainable restaurant, the role of the chef is crucial. The operation of a sustainable kitchen translates into a specific form of organization and management, mainly regarding the sourcing of ingredients. How a chef understands their dishes, ingredients, and the restaurant's relationship with the environment defines the essential features of a sustainable restaurant. Thus, the role of the chef can be understood as that of a social actor incorporated into the social dynamics and actively participating in its construction process.

According to Long (2007), social actors interfere with social reality by remodeling the contexts. Long reiterates Gidden's (1991) concept of "agency" and states that actors are architects of their own actions and of society itself. In this way, society is built through the agency of social actors, who are incorporated into unique contexts and are always creating solutions for their day-to-day dilemmas. In society, there is always room for new creations. The agency allows social structures to reorganize and acquire values and meanings. The agency is characterized by symbolic representations, worldviews, and interests that represent and provide meaning to it. In this study, we understand that the chef's concept of sustainability and sustainable cuisine, which give meaning to this agency, is crucial in analyzing a sustainable restaurant.

How a chef perceives and experiences sustainability in a restaurant is marked by their personal trajectory. As a chef redefines their experience in society, they incorporate their own worldviews, values, habits, and local lifestyles. This is transferred, in a very specific way, to their role and how they create their dishes. In this way, chefs use their profession to express a broader and more generalized social mindset of the community where they live and can be understood to mirror important elements of a widespread mindset in the same community. Therefore, we understand that, individually, a chef conveys the most general characteristics of the community concept of sustainability, expressing the understanding of sustainability embedded in the local culture. All of these meanings, worldviews, and lifestyles are essential subjective aspects in understanding the meaning of sustainability on a local level.

In practical terms, the concept of sustainability gives rise to an agency that reconfigures the restaurant's operations when its chef decides to buy local organic ingredients. Although a restaurant with only local organic ingredients is a yet-to-be-achieved reality, the purchase decision is a pro-sustainability agency. In other words, we understand that the chefs' personal experience and professional performance lead to purchasing decisions that play an important role in managing a sustainable restaurant that is part of the social context, contributing to the creation of more sustainable food production and consumption relationships.

Methodology

To carry out this analysis, we chose the province of Barcelona, located in the Autonomous Community of Catalonia, Spain. Historically, in the city's commercial activities, businesses have represented an important identity trait, and innovation has been a remarkable characteristic. As one of the most-visited tourist capitals in the world, Barcelona has a powerful restaurant industry. Also, the city is home to many restaurants whose owners have a genuine concern for using local and seasonal products. A decisive aspect that explains the existence of such restaurants is the large number of suppliers operating in the region who can offer these products.

In addition to restaurant owners who adopt the slow-food style, meaning they care about the origin of ingredients, sustainable farming, and social justice, a culture aligned with these values has been gaining strength in Barcelona, which can be seen among citizens and government institutions. Civil society has built community gardens, and there is a professional forum on sustainable cooking, among other initiatives. Barcelona signed the Milan Pact in 2015[1] and in 2021 it was the world capital of sustainable food.

Restaurants were selected based on face-to-face and virtual field research and analysis of documents and information collected on the Internet. In Barcelona, the researcher attended training and events on sustainable restaurants offered by the Association of Sustainable Restaurants, which included a wide range of professionals such as nutritionists, farmers, advertising agencies, sales, etc. Finally, in addition to the above, to select the restaurants that would be included in the research, the Slow Food (2020) guide was used to access a list of restaurants that follow the concept of slow food in Barcelona. Based on this, we selected 4 restaurants at random out of the 30 that were awarded Slow Food's Snail of Approval in 2020. The restaurants awarded this seal use locally sourced and handcrafted ingredients, in addition to being part of the area's culture. Three of the restaurants are located in the Barceloneta neighborhood in downtown Barcelona, and one on the outskirts of Barcelona, in the Baix Llobregat region, where many local producers are based. This restaurant was created with the aim of sourcing ingredients from local farmers. We visited each of these restaurants and conducted open interviews with the chefs responsible for the menu and with some of the suppliers.[2] Additionally, during the field research period, the author was in direct

---

[1] The Milan Pact is an international agreement supported by FAO (UN's Food and Agriculture Organization) in which 200 participating cities commit to developing sustainable, fair, and healthy food models.

[2] The issues concerning the relationship between the restaurant and the suppliers go beyond the scope of the present discussion; thus this aspect will be analyzed in another article.

contact with the chefs and suppliers through WhatsApp. We performed a closer analysis of the work environment for the chefs and some of their suppliers at the selected restaurants.

This study understands that each chef, in their own way, has a unique ethnographic trajectory that includes common issues, such as the reinterpretation of cooking, ethnic-identity experiences, values, worldviews, events, and important personal issues, and that they express themselves in their professions, creating particular contexts for their restaurants. Therefore, we sought to analyze the ethnographic trajectory of these chefs who incorporate meaningful family and local elements in their perceptions of food sustainability, creation of their dishes, and management of their restaurants.

Catalan cuisine: cultural identity and sustainability

Ethnic identity is a decisive aspect of strengthening sustainable food practices. In sustainability processes, identity is a motivation. A chef who identifies with the local cuisine and its values, stories, and traditions will feel more motivated to preserve it. Food is one aspect of a society's ethnicity and identity. Food generates an important sense of belonging, which encourages people to carry out actions respecting and caring for the environment, which are the key ingredients in sustainability.

Barcelona is a particularly interesting city to analyze how identity combines with sustainability concerns and how this is reflected in its dishes. The coexistence of cultural references from different regions, due to major migration to the city, alongside its strong culture that seeks to preserve the country's identity, creates an environment where chefs can live rich experiences of innovative gastronomic creation. In Barcelona, chefs are constantly moving through multicultural spaces, with a large number of inputs in their daily lives, which may include new ingredients from other countries that can be grown locally, preparation techniques, recipes, and food cultures, among others. This leads to the emergence of a unique context in which it is possible to discuss the identity of cuisine based on the arrangements and negotiations that are part of the appropriation of many ethnic influences from the city's primary coexisting cultures.

A constant process of reinterpretation is observed among Barcelona's chefs and their kitchens, where to a certain extent, they are concerned with the conservation and promotion of local identity in the face of cultural diversity; and with sustainable values. According to Vázquez-Medina (2018), "ethnicity emerges as a social construct that can be materialized through how one eats and the meanings attributed to this practice" (p. 212). In Barcelona, there is a constant reinterpretation of the meaning of local food and the experience of eating locally, with greater attention placed on the aspect of sustainability and a greater sense of environmental protection in these experiences. At the same time, important features representing the history of Catalan cuisine are maintained, such as the simplicity and seasonal ingredients traditionally used by its people. Throughout the chefs' stories, it could be observed that they recognize sustainability as a path that can ensure the continuity of their local food culture.

When talking about ingredients they use in their kitchens, all the chefs interviewed mentioned private and personal aspects of their lives, such as their family, their childhood, or their mothers and grandmothers. Social and food relationships are built upon ethnic identity and symbolic values that influence the selection of ingredients. In addition, these chefs demonstrate a worldview built throughout their professional careers as they learned about food, its origins, preparation methods, as well as their social and environmental implications.

Identity and belonging are aspects that can raise awareness in chefs and encourage them to seek information about ingredients and the environmental concerns necessary for sourcing them. This also encourages them to establish fairer trade relationships with producers. Therefore, the chef's values and worldviews are key aspects in the selection of ingredients and, therefore, play an important role in the existence of sustainable restaurants. Chefs interested in the origin of ingredients will know how to understand their kitchens holistically and how the ingredients were produced, which are two closely related aspects.

Each society must find those elements of its culture that converge with the principles of sustainability. An important representative of Catalan cuisine was Ferrán Agulló, a journalist born in the 19th century, who was one of the first to write a Catalan cookbook. According to Salvador Garcia Arbós (2000), Agulló, together with Josep Pla, created a "myth of the Empordà cuisine," which is a synthesis of the Giron and old Catalonia cuisines.

According to Llorenç Torrado, who wrote the presentation of the book *El libro de la cocina catalana* by Agulló (1978), this book was written with a deep ideological and cultural weight, seeking to return to the origins in an attempt to establish a national cuisine, in conjunction with a nationalist policy that also encompassed national art and language.

This journalist left the urban context and traveled extensively into the countryside of Catalonia, looking for traditional recipes directly from fishers, farmers, homemakers, innkeepers, sailors, and hunters. According to Torrado, Agulló intended to restore the honor and simplicity of Catalan cuisine based on genuine dishes. However, this simplicity must have grace, even if it does not follow the complex preparations of French cuisine. Agulló sought to avoid the complications of French cuisine and collected simple, healthy, and cheap recipes. According to Torrado, the origins of this cuisine are "limited by ecological conditions, sometimes by misery and in others, by the fact that the same supplies were not always available."

> Thus, a hunter without any pots cooks a partridge buried in the ground and one who has a vegetable garden can adapt the dishes according to the seasonal products, thus, with few elements and little complication, with ingredients that are not necessarily elevated, but with time (if they are not preserved), the dishes can be elevated, through imagination, to good taste and often refinement ((Agulló, 1978) : 9).

Agulló's book collects typical Catalan recipes such as *sofrito, escudella, arroz a la cazuela, macarrones*, veal with mushrooms, and *fricandó*. It also has recipes for basic Catalan sauces, soups, fish, cod, lamb, pork, dishes with vegetables, mushrooms, desserts, etc. The recipes in this book are characterized by their simplicity and seasonality, two important elements for sustainability and something on the mind of the chefs we interviewed.

The chefs reveal deep connections between their professional practice and their Catalan roots. The chef at restaurant 1, during his three-month-long childhood vacations, used to visit Delta del Ebro with his relatives, an area rich in river, sea, and vegetable products, as well as game meat. "We can find very good ingredients in this area," says the chef. Emotions are clearly expressed in the chef's statement, mainly because he mentions people with whom he has a strong emotional connection, such as a woman he calls his "third grandmother." According to this chef, she cooked exquisitely, made her own oil, and had her own rice paddy. During the interview, it became clear that the simplicity of the grandmother's cooking habits had a strong connection with nature and its cycles and seasons.

The owner of restaurant 2 is a Catalan from Barcelona, and he spent his holidays in the Empordà and Girona region, stating that, in his restaurant, some dishes are the same as those "that were served when we (they) were little"[3]. Chef 2, who provided consultancy services for that restaurant, is a Catalan born in a town in the province of Lleida. Tapping into his family influences and local culture, he had two ideas: open a haute cuisine restaurant based on the roots of food culture and develop a creative Catalan cuisine. After that, he opened a restaurant consulting firm. The chef from restaurant 3 is originally from Vilassar de Mar, a town located in the north of the province of Barcelona. This chef tells that since his childhood he liked to eat good things, and by good he understands local fresh food with little or no processing and cooked by his relatives. He talks about how, at his home, there have always been nice meals, especially because of his grandmother, who is a great cook and buys quality products. The chef from restaurant 4 is originally from Esplugues de Llobregat, a town located in Barcelona. His family came from Vilanova de Meià, in the province of Lleida, a town located in the Catalan Pre-Pyrenees. This chef also spent his childhood summers and holidays with his grandparents in this region, and there he always felt connected to nature.

The chefs' stories show that their Catalan roots are a crucial subjective factor that encourages and drives them to create dishes with local products, promoting and strengthening the local economy and lifestyle. There is a general understanding that the path toward sustainability is one that allows for the preservation of local communities—their traditional ways of producing, eating, and living.

Seasonality, product awareness, and cooperation with local producers

Understanding that the path toward sustainability allows for the preservation of local life through its methods of producing, eating, and living encourages chefs to learn more about local products and their producers, as well as to build business relationships with them.

---

[3] Interview with the author.

A distinctive feature of the kitchen of chef 1 is the possibility of working with quality products, which, according to him, are fresh, seasonal, and local with good texture and unique organoleptic characteristics. The worldview of chef 1 was important for him to define the type of business he wanted to open since he was very clear that he wanted to work with local produce, especially during the artichoke season, and cook whatever he could source from the local producers around him. He wanted to work with fresh products; for him the local economy was important, and he wanted to support his town and do something that the locals would visit frequently. Here, it can be observed that the purpose of the business in the chef's life strongly influences his management (agency) as a social actor who participates in Barcelona's restaurant sector with sustainable aims.

All these precepts, values, and attitudes give their restaurant the characteristics that transcend the idea of a merely commercial business. From its very conception, their business proposal reveals a worldview that longs for more than just profit.

According to chef 2, "You have to know who is behind each product, who is growing each thing, visit the farm".[4] There is a connection between ingredients and their history, their origin, and the place from where they are sourced. In his words, "for me, when I receive any product and I know who it came from, it makes me wonder what the producer is like, the landscape they live in, which season we are in, when I'm about to create a dish"[5]. It can be observed that the work of a chef who is focused on sustainability transcends the kitchen. They have to leave their professional environment and go out to the field to learn about the reality of production by talking to producers and understanding the process of growing food. This is all part of the chef's agency, changing the social scenario where relationships are built permeating the restaurants.

The sustainable agency is also observed in a project between Restaurant 3 and a local farmer. The project consists of maintaining a restaurant-owned garden with an exclusively dedicated professional. That was a major change experienced by chef 3: worrying not only about sourcing good produce, but also about maintaining a close relationship with the producer. Chef 3 defines what they do in the restaurant as "a product of proximity, ecological, biodynamic, natural, a result of working

---

[4] Interview with the author.
[5] Interview with the author.

closely with small-scale producers."[6] This is in line with the slow food concept, which, according to him, is what most defines them.

Chef 4 was concerned about knowing where his ingredients came from; he wanted to know who raised the lamb, who made the sausages, etc. So he started to look for the best suppliers in Catalonia. He states that it was not an easy task since it would be simpler to get two or three phone numbers and centralize everything, but he wanted to go further and understand production conditions, the producer's income, the quality of the product, and so on. Those are issues that have always concerned him. This chef explains that, at first, the logistics were complicated because deliveries were difficult, and producers did not want to travel to Barcelona for small amounts, for example, to deliver two blocks of cheese. To solve the problem, he encouraged producers to coordinate among themselves to organize deliveries. That is how they managed to arrange the transportation to Barcelona. Logistics became easier as they acquired more customers in the capital. Here it can be observed that the chef's agency strengthened logistics and sustainable local food trade in conjunction with their suppliers.

Barcelona as a cosmopolitan city: giving a new meaning to Catalan cuisine

Innovation is usually something constant in the restaurant industry, although in some cases, tradition is the main framework. Barcelona is a city where tradition and various referential foreign food cultures intersect, and chefs are immersed in this context. Among all the chefs we interviewed, one could observe that, throughout their lives, their experiences have reshaped their Catalan ethnic identity, and they have embedded elements of foreign cultures and sustainable values that give new meaning to their cooking.

The strong ties, emotional and identitarian, of chef 1 with the products from his region work as mechanisms for him to resignify his kitchen. This chef uses high-quality local products to reinterpret dishes from international cuisine and give them a Catalan touch that represents the local identity. For example, he achieves a fusion of sorts with

---

[6] Interview with the author.

traditional French cuisine, changing the crown roast of lamb for tuna. To do this, he removes the ribs from the tuna, as well as the crowns, and makes a bone stock (which traditionally would be done with lamb bones). In this example, it can be seen that the primary element featured by the dish represents the regional identity, reinterpreting a foreign cuisine and incorporating it into his menu.

Restaurant 2 offers traditional Catalan cuisine, but even in this case, one can observe an element of resignification, which is the idea of an ecologically friendly and sustainable product, which is a modern food concept. The typical Catalan cuisine is served in the restaurant, but with an eco-friendly approach. They understand that they can achieve an authentic Catalan product while following sustainable production standards in harmony with the farmer's culture. The owner's vision of local and natural food translates into organic certification standards and sustainability principles. The owner expresses an ethical approach in the way he runs his business. Sustainability appears as a path to ensure, protect, and preserve the cuisine of all times.

Cooking techniques constitute an important element of food resignification. Chef 4 states that he does not overlook any technique that he finds interesting. For example, he makes vegetable ceviche and explains: "I took the ceviche technique and used it with raw vegetables, Why not? Is it forbidden somewhere? Where are the limits in the creation and use of techniques?"[7]—He asks arguing for his freedom in culinary creation. The chef also states that this is still a Catalan dish because he is using organic vegetables grown 100 km from his neighborhood, and uses a technique that has been used in Catalonia for years. He argues that Peruvians have been living in Catalonia for about 12 years, and the Catalan culture has absorbed their dish. Although ceviche is not a traditional dish, he affirms that it is "already part of the gastronomic ideology that can be found in Catalonia, without a doubt"[8]. The vegetable ceviche is an example of incorporating a foreign dish into Catalan culture through substitutions and negotiations that give a new meaning to the Peruvian ceviche. The chef prepares a Catalan-style ceviche because he uses locally grown products. His standing, as well as that of the other chefs, reaffirms the legitimacy system that resignifies the identity of Catalan cuisine.

[7] Interview with the author.
[8] Interview with the author.

Chef 2 thinks that foreign cooking influences can intermingle, and he does not see any loss of identity or homogenization in the globalization of cuisines. He wonders, "Why not make a Peruvian ceviche with local products?" "Instead of using lime, why not use a lemon from the Ebro Delta?" He continues, "use a technique to intermingle Catalan products and Catalan culture, adding value to the dish"[9]. Here, we see the same argument again, which is part of the negotiations and substitutions that resignify local food processes. Chef 2 does not believe anything foreign should be prohibited but rather used to add value to local products and producers. He also states: "If you love Asian cuisine, why can't you make Asian cuisine with Catalan products and thus transform Asian cuisine into Catalan cuisine?"[10]

The proposal to replace foreign ingredients with local ones and thus achieve a Catalan-style dish can be understood as a way of legitimizing the cuisine created by different chefs, which, in the future, may give rise to a quite varied Catalan cuisine, yet marked by local and handcrafted products. It can be observed that, in Catalan ideology, products play the role of "guardians of the authenticity of Catalan cuisine."

Conclusions

In Barcelona, the perception of sustainability lies in the social context, rich in multicultural spaces, while preserving a Catalan culture deeply rooted in its origins and familial experiences in its villages, keeping it simple, close to nature, and always respecting seasonality. Added to this, there is a genuine concern for sustainability. The local product is the key element that appears as the "sustainable solution" that can preserve the local culture, its suppliers, the environment, and the economy, and also serve as an important element in the resignification of traditional and foreign dishes that are increasingly more present in the city. It conveys a product-based idea of identity and sustainability.

In this respect, chefs identify with the menus they create while feeling free to fully exploit the diverse cultural references with which they are constantly in contact.

[9] Interview with the author.
[10] Interview with the author.

A subjective element is observed in the chefs' statements, as they clearly express their affective ties and respect for the food experiences they had in their families. They recognize that in their products, they use a series of material and symbolic issues that point to their family experiences in their childhood.

Likewise, their kitchens employ dynamic strategies that resignify traditional dishes and negotiate transformations, for example, when they use an Asian technique with local products and claim that Catalan identity is present in the result, affirming that the use of local product ensures its "Catalanity." The process of giving new meaning to the cuisine is a complex phenomenon marked by disruptions, innovations, and replacements. It also involves the creation of a system to legitimize these new eating habits.

We understand that the chefs we interviewed do not represent the entirety of Barcelona's local cuisine. However, they express a strong local mindset trend when they speak of locally produced seasonal products grown in the most natural way possible, which ensures local identity, and thus they understand the generalized local concept of sustainability. This concept of sustainability gives meaning to the chefs' agency and their restaurants, favoring sustainable entrepreneurship with respect and a positive appreciation of suppliers and their products. Therefore, we understand that these restaurants, although they do not meet all the requirements that characterize a sustainable restaurant, play an important role in the initial process of establishing this type of cuisine.

References

Agulló, F. (1978). *Llibre de la cuina catalana* (2nd ed.). Barcelona: Alta Fulla.

Giddens, A. (1991). *As consequências da modernidade*. São Paulo: Unesp.

Garcia Arbós, S. (2000). De les receptes familiars a la gastronomia creative. In *La cuina. Revista de Girona*. n. 200 maig-juny; pp. 363–374.

Goodman, D. (1999). Agro-food studies in the 'age of ecology': nature, corporeality, biopolitics. In *Sociologia Ruralis: European Society for Rural Sociology*, Oxford, v. 39, n. 1, pp. 17–38, jan.

Guivant, J. (2002). Riscos alimentares: Novos desafios para a sociologia ambiental e a teoria social. *Desenvolvimento e Meio Ambiente*. n.5, 2o semestre.

Long, N. (2007). *Sociología del desarrollo: una perspectiva centrada en el actor*. México: Centro de Investigaciones y Estudios Superiores en Antropología Social.

Murdoch, J., & Miele, M. (1999). 'Back to nature': changing 'word of production' in the food sector. In *Sociologia Ruralis: European Society for Rural Sociology*, Oxford, v. 39, n. 4, out.

Nascimento, M. (2014). *A. As práticas alimentares na sociedade globalizada: O caso do movimento Slow Food*. Florianópolis, Brasil: Tesis y Trabajos. UFSC.

Oosterver, P., Guivant, J., & Spaargaren, G. (2010). Alimentos verdes em supermercados globalizados: uma agenda teórico-metodológica. In J. Guivant, G. Spaargaren, & C. Rial (Eds.), *Novas práticas alimentares no mercado global* (pp. 15–57). Florianópolis: Ed. da UFSC. (orgs.).

Slow Food. (2020). Barcelona slow food guide. Slow food Barcelona.

Spaargaren, G., Oosterveer, P., & Loeber, A. (2012). *Food practices in transition: Changing food consumption, retail and production in the age of reflexive modernity*. New York: Routledge.

Vázquez-Medina, J. A. (2018). Alimentación, etnicidad y segunda generación en un fenómeno de migración internacional: Ser mexicano y comer mexicano en Estados Unidos. In M. Bertran, & J. A. Vázquez-Medina (Eds.), *Modernidad a la carta. Manifestaciones locales de la globalización alimentaria en Mexico*. Icaria: Barcelona.

Watts, M., & Goodman, D. (1997). *Globalizing food. Agrarian questions and global restructuring*. London and New York: Routledge.

CHAPTER EIGHT

The cuisine of the new spirit of capitalism: Noma considerations regarding the value of the authentic and other orders of worth

Joan Frigolé
Department of Social Anthropology, Faculty of Geography and History, Universitat de Barcelona, Spain

Introduction

Noma is a Danish restaurant in the Nordatlantens Brygge building, an old wine cellar in Copenhagen now converted into a cultural center of the North Atlantic area. Considered the "Best Restaurant in the World" by the *Restaurant Magazine* guide in 2010, 2011, 2012, 2014, and 2021, it also has three Michelin stars. Its dishes are based on local ingredients represent the international image of modern Danish cuisine. René Redzepi is the head chef and coowner of the restauran t.

Noma has recently undergone a significant transformation. The Danish restaurant not only changes its menus with the season, but also transforms its context, façade, landscape, and appearance three seasons a year. The new Noma (or Noma 2.0) started in February 2018, a year after closing at 93 Strand Street, where it was originally founded in 2003. There was a reason for closing its premises on the wharf in the Christianshavn district: to move to a new site where the project could grow not just in size, but also in culinary freedom and ambition.

Our initial question: What does Noma reveal about modern capitalism? Noma, an acronym for the Danish term *nordik mad*, meaning Nordic food, was the name given to the restaurant in 2010, and since then it has held the first place in the world restaurant rankings on several occasions, while its head chef, René Redzepi, aged 44, is considered one of the finest chefs in the world. This is a success story, and is thus famous, and this is used here to illustrate a number of concepts presented below and certain changes that have occurred in the capitalist system.

It should be noted that this article does not focus on the latest stage of Noma at its new site, but mainly on its previous history, which may be considered foundational and programmatic. Thus the text consists of three parts: the theoretical framework and main concepts, discussion of the empirical case, and a final brief contextualization with references to other cases and factors that characterize current reality.

Theoretical approach and main concepts

The analysis I intend to conduct here is an extension of a previous one (2010) that defined and related the concepts of patrimonialization and commodification of the authentic. My aim here is to place these concepts in the general theoretical framework provided by the models of "orders of worth" and "spirit of capitalism," developed respectively by Boltanski and Thévenot (1991, 1999, 2000) and Boltanski and Chiapello (2002, 2005).

The sociology of Boltanski et al. focuses particularly on critique and justification, as well as the notions of justice and injustice formulated by people in situations and at moments that are critical to them and society. The notion of critique and justification is central to the spirit of capitalism model. The spirit of capitalism arises from considering "the interaction between capitalism and its critics" (2005: 240), i.e., it captures the need to justify capitalism "simply because it is the object of criticism" (2005: 241−242). Boltanski and Chiapello define the spirit of capitalism in the following terms: "The spirit of capitalism is precisely the set of beliefs associated with the capitalist order that helps to justify this order and, by legitimating them, to sustain the forms of action and predispositions compatible with it. These justifications, whether general or practical, local or global, expressed in terms of virtue or justice, support the performance of more or less unpleasant tasks and, more generally, adhesion to a lifestyle conducive to the capitalist order." (2002: 46)

Boltanski and Chiapello consider "the concept of a spirit of capitalism allows us to combine within one and the same dynamic the changes in capitalism and the critique that has been made of it (...) The critique is a catalyst for change in the spirit of capitalism." (2005: 242) They identify two types of critique of capitalism formulated since the 19th century: the "social critique" and the "artistic critique." They characterize the artistic critique as: "This form first emerged in small artistic and intellectual circles, and stresses other characteristics of capitalism. In a capitalist world, it criticizes oppression (market domination, factory discipline), the massification of society, standardization, and pervasive commodification. It vindicates an ideal of liberation and/or of individual autonomy, singularity, and authenticity" (2005: 242). In their opinion, capitalism prioritized the response to the "social critique" and ignored the "artistic critique," but later incorporated part of the content of the "artistic critique" into a justification of capitalism, in what they term the third spirit of capitalism and which they situate in the decade of the 1980s, taking the development of capitalism in France as a reference. They construct a typology of the "spirit of capitalism" combining four criteria: the form adopted by the capital accumulation process; the stimulations of capitalism; fairness following rules that are made public; and the type of security it provides (Table 8.1) is a reproduction of the characterization of the third spirit of capitalism.

The concept of the test is central to this model of change in capitalism. The term refers to subjecting something to a test to check its quality or value. Doing so first requires a choice of the convention of equivalence, such as creativity, productivity, or some other. Conventions of equivalence refer to orders of worth that the test translates into terms of merit. Part of the critique refers to the design of the tests and their performance. The concept of test constructed by Boltanski and Chiapello also includes the concept of the trial of strength as defined by Latour (1984), which is associated with the concept of the social network. Trials of strength

Table 8.1 Reproduction of the characterization of the third spirit of capitalism (Boltanski & Chiapello, 2005: 245).

Forms of capital	Network companies, Internet and biotechnology, *accumulation process* global finance, varied and differentiated production.
Stimulation	No more authoritarian bosses, organizations with fuzzy forms and limits, innovation and creativity; permanent change.
Fairness	New form of meritocracy that values mobility, the ability to nourish networks, each project is an opportunity to develop employability.
Security	For mobile and adaptable employees, companies provide self-help resources to manage oneself.

constitute "moments of confrontation that are not institutionalized, controlled, codified or regulated, which nevertheless produce transformations in the confronted bodies" (Boltanski & Chiapello, 2005: 259).

Previously, Boltanski and Thévenot, in their book *De la justification. Les économies de la grandeur* (1991) formulated the analysis model of "conventions or orders of worth." They state: "the main objective of the model was to provide an instrument with which to analyze the operations people perform when they use criticism, have to justify the criticism they produce, justify themselves when faced with the criticism they receive or collaborate in the search for a justified agreement. The privileged object of the model is made up of situations that are subjected to the *justification imperative*." (2000: 208−209)

The term orders of worth refers to the justifications social actors use to base their position in situations marked by dispute and disagreement. Justifications refer to legitimate values from which people's things and actions are valued and qualified. Boltanski and Thévenot identify six historically constructed orders of worth. These orders of worth are specified in six conventions with their respective modes of evaluation: Market (price and economic worth), Civic (collective interest), Domestic (esteem), Public (view), Inspiration (creativity), and Industrial (Productivity).

The typology of orders of worth uses the following criteria: mode of evaluation, form of proof, elemental relation, and human rating or quality (Table 8.2). shows the three orders of worth and their characterization based on a table by the authors (1999).

The models of Boltanski et al. do not reduce social relations to power relations or people's strategies to optimize their interests, as occurs in various forms of sociology derived from utilitarianism. Referring to the latter, Boltanski and Thévenot write, "These constructions do not consider the demands of justice that people express and reduce them to ideological masks or ignore them." (2000: 209)

Table 8.2 Convention theory: one of the main contributers to the conceptualization of quality (Barham, 2003: 130).
Orders of worth

	Domestic	Civic	Commercial
Mode of evaluation	*Esteem*	*Collective Welfare*	*Price*
Format of relevant proof	*Oral*	*Formal*	*Monetary*
Elemental relation	*Trust*	*Solidarity*	*Exchange*
Human qualification	*Authority*	*Equality*	*Desire, purchasing power*

According to Boltanski and Chiapello:

Awareness of the constraints and opportunities people have to face is not enough to understand what people do; the reasons by which people justify their actions also need attention. Pure calculation of interest or attributing everything to social determinism are just two ways of denying the importance of interpreting what people do, the criticism and distancing people are capable of in relation to their experiences, and the reflection, particularly on moral problems, they undertake. (2005: 240)

Finally, I focus on the concepts of patrimonialization and commodification of the authentic. The latter comes from Boltanski and Chiapello's book, *The New Spirit of Capitalism*: "The commodification of the authentic thus assumes reference to an original that is not a commodity good, but a pure use–value defined in a unique relationship to a use" (2002: 559). It "...consists in exploiting goods, values and means under the sway of capital which, while acknowledged to represent riches (or "treasures") (...) were nevertheless hitherto excluded from the sphere of capital and commodity circulation." (2002, 559)

The commodification of the authentic operates the "transformation of noncapital into capital" through "a series of operations that may be called operations of production, since they have the effect of creating a 'product' out of diverse resources." (2002, 560). These operations are:

1. "Exploring sources of authenticity."
2. "Analyzing the product to control its circulation and make it a source of profit."
3. Subjecting it to "an operation selecting the distinctive features to be preserved," in other words, "codification." (Boltanski & Chiapello, 2002: 560–561)

There is a difference between codification and standardization. "Standardization consisted in conceiving a product from the outset, and reproducing it identically in as many copies as the market could absorb," while "codification, element by element, makes it possible to operate on a combinatory, and introduce variations in such a way as to obtain products that are relatively different, but of the same style. In this sense, codification allows for a commodification of difference that was not possible in the case of standardized production." (Boltanski & Chiapello, 2002: 561)

I base the presentation of patrimonialization on a previous text (Frigolé, 2010). Kirshenblatt and Gimblett define the term heritage as "a cultural production in the present which has recourse to the past" (2001: 44). Guillaume, who uses the equivalent term conservation, defines it as "the representation and materialization of the past in the present and for the future" (1990: 15).

The reference to the past is central in both definitions. Patrimonialization refers to the heritage production process. Production is both material and symbolic. The material dimension is more visible when a building, landscape, or object is being reconstructed or restored. It is always symbolic. Patrimonialization converts a patch of territory into a protected area and a building into a monument. This change implies a de-contextualization, that is to say, a physical and symbolic separation from other elements in its surroundings, and then a re-contextualization. Patrimonialization entails the existence of a space differentiated by objects and other patrimonialized elements, usually a museum, a natural park, or a similar space, and if that is not possible, differentiated times. The exhibition function usually blocks out other functions and makes it clear that we are dealing with heritage; it is made visible and visitable. Cultural production is superimposed upon the initial, immediate, and primary production of an object or element. The relation to the past is complex, for this is not a given fact or reality. Production of heritage also entails the production of the past, that is to say, the selection and manipulation of the past.

Patrimonialization and commodification of the authentic complement each other in the analysis of a tertiary economy. The main correlations between both concepts are:

1. The idea of production. They are processes that produce value, heritage value, and the authentic.
2. The reference to the past, conceived either as a start (culture) or origin (nature) is fundamental to the definition of patrimonialization. In the definition of the commodification of the authentic, the reference to the past is deduced from the characterization of the objects included in it as those that "were still outside the sphere of capital and mercantile circulation," that is to say, outside the development of capitalism identified as historical development.

I will now relate both concepts to the orders of worth approach. A basic idea: the reference to orders of worth precedes the qualification operations, i.e., attributing worth to subjects, actions, and objects.

Given that the past as memory, tradition, or origin has become a fundamental reference for our time and our society, I consider a further order of worth that can be added to the Boltanski and Thévenot typology, which may be termed heritage. Its mode of evaluation would be reference or fidelity to the past. Patrimonialization fully matches this order of worth, while commodification of the authentic does so only partially, in that it refers to the charm of elements that have been left behind by historical development and which may relate more closely to the order of worth centered on inspiration and creativity, which fosters attitudes of rejection toward the conventional, stereotypical, or standardized.

The heritage order of worth reflects a fascination with the past and considers it a source of value. In its current form, this order is not old. An indicator of such is the fact that 1980 was designated "heritage year" by the President of France, the first example of such an initiative. (Barham, 2003) The rapid acceptance of this order, as indicated by the numerous heritage institutions and the fact that the term is applicable to almost everything (the former president Lula da Silva described the EU as world heritage) can be related to numerous factors, such as the reaction to increasingly rapid processes of modernization followed by globalization, concern over food safety, and the dissemination of the concept of biodiversity with its central opposition between indigenous and foreign, populism, and nationalism.

In short, the commodification of the authentic and patrimonialization play a significant role in modern capitalism because they connect with orders of worth that provide ideas and values, such as creativity and authenticity, which the third spirit of capitalism incorporates as justification against criticism and because they produce new types of values that boost capital accumulation, the basis for the development of capitalism.

Noma: a model of gastronomic excellence

The information in the example comes from three long articles that include narratives and statements from the protagonists. This data is sufficient, given that my aim is not so much to conduct an in-depth analysis of the case, but to suggest that the theoretical approach and main concepts discussed here are relevant to its analysis.

The presentation of the case is broken down into six main aspects: the idea of the test; the domain of the wild; the wild as an icon of the indigenous; the architectural and institutional context; labor relations; and cooks and customers without intermediaries.

The idea of the test

Noma gained its world number one ranking in 2010. A journalist writes: "It was the first time in five years that the first two places were not taken by Ferran Adrià's El Bulli and Heston Blumenthal's The Fat Duck. Furthermore, Noma had overtaken Mugaritz, near San Sebastian, the restaurant that experts rumored would obtain a third Michelin star" (Steingarten, 2011: 246).

This is not a one-off or exceptional test but one of daily experiences conceived as a challenge or test within the daily experience. René Redzepi, the head chef, expresses it like this: "Each day is like the Champions League final. You are playing a crucial match twice a day for 80 people who have been waiting months for a table... If they score there's nothing you can do. I don't accept mistakes. It would mean betraying these people who give us their time and money." (Rodríguez, 2011: 35). He uses a sporting metaphor that condenses the ideas of maximum competition and tension to classify the daily challenge, to which the words "passion and singular dedication" are a response, as highlighted by the journalist. It is the quality and value test when faced with expert and nonexpert diners, in which you need to show just how much "innovation and creativity" you are capable of.

The domain of the wild

I have reconstructed the narrative of a pathway using information from two articles that express a basic similarity:

We drove to a beach an hour to the west of Copenhagen, where we walked along the muddy sands and among the low scrub surrounding it, gazing at the profusion of wild plants under our feet. Every so often, René bent down to pick a leaf and pass it to me. Have you ever tried shore dock? (...) And when was the last time you chewed a bulrush? (its flavor is a mixture of cucumber and pepper, fresh and crunchy, which appears after cutting off several woody layers). (Steingarten, 2011: 246).

This beach, on the Dragor coast, is defined "as one of the wild pantries that feeds Noma," and where, as René informs the journalist:

Everything is food, food, food. Careful not to step on it! This territory is full of life and flavor. (Rodríguez, 2011, 41)

We then go to the nearby Brogard farm. (...) To be honest, I wasn't sure if we were stepping on crops or fallow land, where a multitude of herbs and yellow spring flowers had been left to grow wild. (...) A parcel at the start did not seem to have been planted, until I realized there were heads of white asparagus sticking out from the brush. (Steingarten, 2011: 246)

Next we went to a forest of perennial trees. René pinched the new light-green needle shoots growing at the end of all the branches, and gave them to us to taste. They were surprisingly edible, sweet and tender. (Steingarten, 2011: 248)

The walk around the natural and rural environment starts on a muddy beach, moves on to a farm where crops blend with the predominant brush, and ends in a forest, the epitome of the domain of the wild. The journalist's focus of attention on the restaurant and its cuisine is displaced toward the wild environment, as reflected in a number of expressions: small wild plants, wild pantries, running wild. René Redzepi gets the journalist to experience a model journey reminiscent of his own initial trip that resulted in the radical change to his gastronomic model. The number one chef defines himself and his 25 cooks as: "gastronomic explorers" (Rodríguez, 2011: 34) and projects an image of a naturalist and gatherer rather than a cook: "for me, seeing where and how a berry grows or a bee pollinates a flower that is going to be used in the same dish is more important than a kilo of Iranian caviare." (Rodríguez, 2011: 35). The importance of gathering is conveyed by the following information: "We gather 1500 kilos of plants for the winter." (Rodríguez, 2011, 35)

The journey through the natural and rural environment described here could be considered a ritual, recalling the initial journeys made by the chef and his partner. Before opening the restaurant, they both took a trip to the North Atlantic territories:

We discovered produce we had never tried before: turnips, cold-water fish, rye bread, dairy products, cereals, ancient crustaceans, goat and caribou meat... food that grows clean and free, which nobody uses and which had disappeared from Nordic cuisine. We were astonished by these flavors. But it was easier for a Danish cook to obtain French foie-gras than these products. They could not be obtained at any price. This was our first investment. (Rodríguez, 2011: 40)

The narrative of a second journey: "In March 2004, in the midst of an existential crisis, with the restaurant empty, they returned to Greenland. (. . .) When they were about to return to Denmark, they got stuck in the only airport in Greenland, an old cold war military base, due to an apocalyptic snowstorm. 'We were cut off there for a week. (. . .) It was a magical moment. I realized that I had to put this immense natural world in which we were submerged to the service of our diners. (. . .) Each dish had to be surrounded by its habitat and clean and simple in its complexity.'" (Rodríguez, 2011: 40).

The stories associated with the marketing of the authentic are stories of initiative, discovery, intuition, adventure, and so on. Displacement is an important factor. In contrast, it provides a vision that encompasses an assessment of landscapes, habitats, production systems, lifestyles. In short, a vision of societies and cultures different from one's own, with their own specific values. This displacement includes travelers from distant lands and those who explore remote areas of their own country or settle in them, as in the case of neo-ruralism. Discovering, identifying, and seeing something no one else sees or values, as Redzepi's warning to the journalist highlights: "Careful where you step, this is food, food, food."

The spirit of discovery and all the paraphernalia that surrounds it (for instance, consider the journeys to the Andes or remote parts of Asia by certain business people) is essential for the finding of unconventional "riches" that can be marketed under the label of the authentic, which refers to the indigenous and territorial. The value of these "riches" lies not so much in their bonds with a territory but in the possibility of converting them into exchange values and setting a price for them.

It is not enough to discover new "sites" of gastronomic interest; a fairly continuous and regular supply must also be ensured, given that they are not initially goods, or if they are, they are part of an irregular and distant market. This gastronomic model reinforces the importance of the local gatherer, not as an archaic figure but as a fully modern one.

The return to nature, a nature unknown or forgotten in gastronomic terms, is the basic premise of the Noma gastronomic model. René Redzepi attributes two basic characteristics to nature: (1) it provides an "amazing pantry"; and (2) "it gives you a reference for how natural things should taste; how Noma's dishes should taste." (Rodríguez, 2011: 41).

Nevertheless, the criteria for codifying his gastronomic offer do not come directly from nature, but from a prior codification of nature that is part of the Nordic design (already a brand) and its transfer to gastronomy. The characteristics of the Nordic design brand are quality of life, simplicity, cleanliness, naturalness, functionality, purity, freshness, sobriety, elegance, respect for the environment, and harmony with nature. Similarly, Noma's food "is subtle, colorful, fat-free, gives the impression of having barely been touched in the kitchen. You eat it with your fingers. Rawness predominates. There are no portentous or aggressive flavors, no big explosions of flavor. (. . .) It is gastronomy that tastes of land, sea and forest." (Rodríguez, 2011: 42) This gastronomy is largely the result of the transposition or transfer of essential characteristics, meanings, and values from another area of culture.

The wild as an icon of the indigenous

The emphasis on the wild reinforces the value of the indigenous and the territory. This is clearly expressed in one of the texts: "Shore dock or forest grass; if it's indigenous, René Redzepi can use it." (Steingarten, 2011: 246)

As a restaurant, Noma is characterized by its use of local ingredients. The journalist specifies the meaning of the term local: "I mean from Denmark, Norway, Sweden, Greenland, Iceland and the Faroe Islands. Noma was going to be a restaurant by and for "locavores" (a word coined in 2005 by four women in San Francisco, which in 2007 the New Oxford American Dictionary named word of the year). (...) Undoubtedly other chefs throughout Europe and America have started out on a similar journey, but none has reached the same place as René. The first step in eating like a "locavore" is to introduce local ingredients into the predominant cuisine, and eliminating anything that does not grow in the region. (...) No tomatoes, because when they are from Denmark they have little gastronomic value. No aubergine or mango, which do not grow here. René replaced olive oil with hazelnut, beech nut, mustard seed and rapeseed oils. Even here, he went further than most "locavores." For René that was just the start. He might not have been looking for a revolution, but with his unstinting and ingenious exploration of nature, René is creating a cuisine that offers satisfaction like no other." (Steingarten, 2011: 248)

What we have here is a radical interpretation of the indigenous and nonindigenous dichotomy. The chef's creativity is based on the radical nature of this premise and is largely expressed through the principle of substitution. The values of purity and sincerity are central to the justification of this gastronomic proposal, as various journalists stress: "René is a maestro in the art of local food, the most ingenious purist I have known." (Steingarten, 2011: 249) "Back to the authentic. (...) His success is based on the cyclical return to evident products and specific and sincere flavors." (Cepeda, 2010)

Valuing the authentic and indigenous can sometimes produce a negative impact on biodiversity. In North America, "the lucrative wild leek market has been attracting hordes of diggers who take unheard-of quantities out of the forest. Several botany experts think this obsession with and glorification of the wild leek has triggered excessive harvesting" (Patil, 2011: 1). Canada has declared it a protected species.

The architectural or institutional context

The building is described as "an ancient and beautiful port warehouse from the 18th century" (Rodríguez, 2011: 40) that underwent restoration. Its interior is described in the following terms: "The light is natural, and the space large and clean; the sparse furniture mixes tradition and Nordic design; the floor is covered by warm Pomeranian pine, and the old 18th-century ceilings are supported on broad oak beams. It has the scent of the village; of wood, wax, cereals, smoked fish and flowers." (Rodríguez, 2011: 42)

The restaurant is part of an institutional complex called the North Atlantic House, which also houses an exhibition hall, a museum, and the permanent representations of Greenland and the Faroe Islands in Copenhagen.

This institutional context is complemented by the Nordic Cuisine Symposium, which Meyer, an entrepreneur, expert in gastronomy, and Redzepi's main partner, organized in 2004, bringing together Scandinavian chefs with the aim of reflecting on and drawing up a common philosophy. "This congress would give birth to the so-called New Nordic Cuisine, a movement that combined the philosophy of the Meyer/Redzepi duo and involved Swedish, Norwegian, Danish and Finnish chefs. A few months later, this cultural-gastronomic-commercial current was ratified and supported by the Nordic Council, a confederate forum with the region's foreign ministers, who saw in this prestigious label an unexpected opportunity to export and attract tourism." (Rodríguez, 2011: 41).

Labor relations

The restaurant has a capacity for 40 diners and an identical number of staff. René refers to the cooks in his restaurant in these terms: "My cooks are under 30, and come from all over the world." (Rodríguez, 2011: 34) According to one journalist, they are 25 and "attractive, young, smiley, multilingual, didactic and tattooed." (Rodríguez, 2011, 42)

A large number of the employees also work in a specialized way, the opposite of standardization, which demands a great deal of work. René describes himself as "I feel more of an artisan than artist." (Rodríguez, 2011: 34). This does not mean the restaurant belongs to the field of craftsmanship.

One journalist observes that René's behavior "transmutes" when he is in the restaurant, and "he can mercilessly lay into any of his 40 employees because of a badly cleaned floor, a creased serviette or a dish out of place." (Rodríguez, 2011: 35) One employee describes René's position, including his authority, in relation to his subordinates in these terms: "René's demands in the kitchen are unhealthy; he puts up with a lot of pressure; not just to make the restaurant work and keep it at the top, he has to create new things and meet the media. When something goes wrong he goes crazy. He's like a volcano. He can crush you." (Rodríguez, 2011, 35) René himself recognizes this: "It's true, I can make a cook cry; it happens every day; the kitchen in a great restaurant is driven by stress, passion and dedication. We are visceral people and sometimes we blow up. We work 14 hours a day with barely a break." (Rodríguez, 2011, 35).

What are these employees paid? One journalist who speaks to some of them writes, "the rookie chef is American and doesn't get paid. He is one of this new generation of cooks without borders who goes from one of the world's great restaurants to another, paying for their future as a global super-chef with their sweat." (Rodríguez, 2011: 35).

Creativity not only has an impact outside the restaurant, i.e., in the market, but also inside, in the workforce, to the extent that it replaces the salary with the motivation and satisfaction it produces. Entering, staying at, and leaving the restaurant successfully is a qualifying test. Future employability or the future condition of an entrepreneur is based on the experience of exploitation.

The modern aspects of this hugely successful capitalist business (in 2010, it received over 100,000 booking requests on its website) are mixed with other aspects we might consider archaic, such as unpaid apprentice work. One does not exist without the other.

One of the features of modernity is the existence of a laboratory as part of the restaurant. As René says: "on Saturdays, when we finish, we stay to try the new dishes the lab team has produced." (Rodríguez, 2011: 35). Science and technology applied to gastronomy indicate the idea of progress linked to capitalism, but the claim of renouncing "advanced technology" must also be considered (Cepeda, 2010). Perhaps this is because, given its association with the production of artificial food, it would contradict his model of indigenous nature-based gastronomy.

Using the opposition between cultural societies and scientific societies as a reference, Martínez de Alberdi states that, in the latter, the market spaces, restaurant environment, kitchen, dining room, management, research lab, etc., create "a structure of communicating vessels, an expert network around which circulates the gastronomic experience as a comprehensive process with an experimental and interdisciplinary vocation." (2008: 272)

Cooks and customers without intermediaries

The direct relationship between the cook and the diner is the general model of the restaurant. René describes his diners and how they are treated: "they are normal people who wait months for a table and who are served personally; I explain my dishes and what I connect with. (...) I can see in real time how people react to what they eat." (Rodríguez, 2011: 34). The explanation and conversation about the dishes help blur the difference between experts and nonexperts. Eliminating intermediaries is not only related to the didactic purpose and control of the culinary process, from its start in the kitchen to its conclusion on the table, but also with the value of contact with the creator, personalized attention, linguistic competence, charm, youth, and beauty. The treatment is part of the product and constitutes its value. The product is often compared to art. One journalist states: "René's cuisine reaches the level of art in the physical and visual composition of his dishes. They are exquisite still lifes. (...) And with each one René Redzepi creates a whole world." (Steingarten, 2011: 249) In one of the restaurants mentioned below, the quality of services has been reinforced by incorporating elements from choreography and dance. The idea is that the diner participates in an artistic performance.

A brief contextualization of Noma

The contextualization has two aspects. First, a partial description of the two restaurants with which Noma shares a "similar family likeness." Second, the presentation of a number of general factors that may help in understanding how Noma has created an aura of exclusivity through creativity applied to the domain of the natural and indigenous.

1. Other similar cases:

The first one is the Swedish restaurant Fäviken. The journalist presents its head chef as follows:

> *The naturalist Magnus Nilssen makes his mark in the new Swedish cuisine." The presence of the term naturalist and absence of the term cook in the title is significant, but this is not a biologist who has become a cook, as one might be led to understand. The description of the restaurant highlights the following elements: "Slate and slabs of granite, together with cut wooden boards, are the basic elements in the Fäviken rustic tableware. (...) The furs from animals that live in the immense forests in northern Sweden and the bark of trees warm the limited clientele. The rural decoration that hangs from the ceiling in some corners: cold meats and cod out to dry. (...) The purity and technical minimalism is the path to "going back to our roots, a legacy that we failed to use to the full." Everything on offer comes from the place or its surroundings. Every day he goes out for a couple of hours to pick plants and shoots. The lichens that cover the stones are served as a crunchy aperitif. (...) He has a garden to grow vegetables organically and also, just by the kitchen, ponds with ducks and fish that end up in the kitchen. (...) He keeps the utensils his grandmother used and even her sourdough, which still provides the slices of bread with their consistency and authentic flavor. The accent on authenticity is so great in the hands and ideas of Nilsson that he himself and a cook from his team saw the bone of the recently slaughtered cow to extract the marrow in front of the public. (Rivas, 2011: 46)*

The second is the Basque restaurant Mugaritz, whose declaration of principles state:

> *From a strict and honest respect for nature, on a clear and precise pathway along the lost trails of the often forgotten flavors and food practices of the Basque homeland, Mugaritz is setting up a gastronomic practice of cooking and presentation capable of producing a discourse beyond the technological, ethereal and textural gibberish of recent times: pursuing the latest artisans, rereading the most adventurous chefs, making the statement of principle that in times of globalization a dish in Biarritz cannot taste the same as in Tokyo. (Martínez de Alberdi, 2008: 279)*

The references to nature and the past, both expressions of the indigenous, stand out in these texts. "A clear and precise pathway along the lost trails of the often forgotten flavors and food practices of the Basque homeland" (Mugaritz) is the equivalent of "Going back to our roots, a legacy that we failed to use to the full" (Fäviken).

Patrimonialization is complemented by commodification of the authentic as "a pure value, defined by a singular relationship with regard to the user" as exemplified by "the utensils and sourdough of his grandmother, which still provides consistency and authentic flavor" to the bread served in the Fäviken restaurant.

2. Some of the factors in the general context of the gastronomic emphases in these restaurants are a new conception of nature, the process of wilding the territory, and the use, not generalized but ever more frequent, of representations of nature by companies in all sectors.

The emphasis on the wild highlights the importance gained from a new concept of nature that differs from the classic one, condensed into the cultivated-domestic/wild opposition. The wild takes pride of place in line with a new conception of nature, perceived and valued as sources of experiences and sensations. It is worth noting that in 2010, when Noma first became world number one, a nonprofit organization was created with the aim of rewilding extensive areas of Europe to restore the wild nature of the past and commodify it as an authentic reality through quality tourism.

Representations of nature linked to notions of biodiversity and sustainability are increasingly used by companies in all sectors associated with their corporate brand to reinforce the perception and appreciation of their products on the market. Sometimes general images are used, such as the following advert: "Your towels plant trees. In this hotel, 5 reused towels = 1 tree planted." The tree has become the supreme icon of nature, and planting trees, thousands or even a million of them, is the challenge that large companies in the consumer sector, such as Walmart and Yves Rocher, have launched at their customers to generate their loyalty. On other occasions the images are of specific elements of nature, such as turtles, dolphins, tigers, eagles, orangutans, crocodiles, or variations of natural spaces. By way of example, see the environmental association of a hotel chain and a beer producer: "Turtle conservation in Palace Resorts. Throughout the sea turtle's nesting season on the Quintana Roo coasts, the three Palace Resorts turtle camps work hard to protect the nests. Since 2006, they have managed to release 496,099 hatchlings. For six months, the members of the camps, headed by biologists and vets, take night walks on the beaches of the hotel zone with the aim of protecting the females and her eggs from predators, including humans. According to the hotel chain director, the campaign also seeks to share this experience with guests, collaborators and visitors, especially children, as a tool for environmental awareness-raising and education." (Anonymous, 2011: 120) The Modelo group, a leader in beer brewing, has become an icon of conservation for its leadership in the conservation program for the Izta-Popo park (Mexico), which receives a million and a half visitors a year.

Many companies do not associate their brand with a specific element of nature but with the central concepts of biodiversity and sustainability, whose translation into accounting and economic language is familiar to people. For instance, an airline company might suggest that passengers financially offset the atmospheric CO_2 emissions their trip generates. The idea of sustainability is accepted by many companies in how it relates to innovation and modernization of productive processes and, therefore, savings, the projection of a positive brand image, added value that distinguishes it, and gives the product competitive excellence. Companies look to turn their logos into eco-symbols.

The widespread dissemination and importance acquired by the notion of biodiversity and its key differentiation between indigenous and foreign have generated powerful natural images often inscribed in demonstrations, experiments, etc., by scientists and experts. These are presented as scientific actions and techniques for selection, purification, protection, and extermination of natural elements, but transcend the scientific-technical level to become metaphors for the social order and its problems. The presentation of these apparently technical actions by the media and the explanations and justifications that the presentation itself generates imbue them with a ritual dimension in that they celebrate and disseminate a natural value that is easily convertible into a social value. Radical love for indigenous nature easily becomes radical love for indigenous culture and society.

This is the context in which Noma was born and is situated.

To conclude, in itself and in terms of its culinary products, Noma is one of the varied and differentiated productions essential for the third spirit of capitalism, in terms of both ideology and justification and in terms of capital accumulation. Varied and differentiated productions are the opposite of standardized and generic productions. These poles of specialization and differentiation are where the productions of patrimonialization and commodification of the authentic are located. Through both mechanisms, Noma has developed innovation and creativity, operating radically with natural and indigenous elements.

Acknowledgments

The idea and a first unpublished version of this text were conceived as an invited conference at the Department of Social Anthropology of the University of Barcelona, in November 2011. Although some parts of the text have been updated to reflect the changes that have occurred up to now, this text. It mainly collects the ideas and the discussion that motivated my reflection at the time. This structure has been maintained in this chapter. I would like to thank Oriol Beltran and Míkel Aramburu, from the Department of Social Anthropology of the University of Barcelona, for their comments and contributions to that first version of this text.

References

Anonymous. (2011). Conservación de tortugas en Palace Ressorts. *MDC Magazine. Mercado de Convenciones, 85,* 120.

Barham, E. (2003). Translating terroir: the global challenge of French AOC labelling. *Journal of Rural Studies, 19,* 127−138.

Boltanski, L., & Chiapello, E. (2002). *El nuevo espíritu del capitalismo* (1st ed., p. 1999) Madrid: Akal.

Boltanski, L., & Chiapello, E. (2005). The role of criticism in the dynamics of capitalism. Social criticism versus artistic criticism. In Max Millar (Ed.), *Worlds of capitalism. Institutions, governance and economic change in the era of globalization* (pp. 237−267). London: Routledge.

Boltanski, L., & Thévenot, L. (1999). The sociology of critical capacity. *European Journal of Social Theory, 2*(3), 359−377.

Boltanski, L., & Thévenot, L. (2000). The reality of moral expectations: a sociology of situated judgement. *Philosophical Explorations, 3,* 208−231.

Cepeda, L. (2010). Noma copenhague. *Paisajes Magazine,* September.

Frigolé, J. (2010). Patrimonialization and the mercantilization of the authentic. Two fundamental strategies in a tertiary economy. In Xavier Roigé, & Joan Frigolé (Eds.), *Constructing cultural and natural heritage. Parks, museums and rural heritage* (pp. 27−38). Girona: Institut Català de Recerca en Patrimoni Cultural (ICRPC).

Martínez de Alberdi, I. (2008). Hacia una reconsideración del patrimonio gastronómico. La experiencia gastronómica en las sociedades de ciencia. In Marcelo Álvarez, & F. Xavier Medina (Eds.), Identidades en el plato. El patrimonio cultural alimentario entre Europa y América (pp. 259−280). Barcelona: Icaria and Observatorio de la Alimentación.

Patil, A. (2011). Vida (comestible) más allá del jardín. *The New York Times- El País,* 1.

Rivas, R. (2011). Delicias vikingas para todos los gustos. *El País,* 46.

Rodríguez, J. (2011). Redzepi. El señor de los fogones. *El País Semanal, 1827*(2), 32−43, October.

Steingarten, J. (2011). El festín de René. *Vogue number, 274,* 246−249, January.

CHAPTER NINE

Food on wheels: culinary paths toward sustainable lives for migrants in Germany

Edda Starck and Raúl Matta

Institute of Cultural Anthropology/European Ethnology, University of Göttingen, Göttingen, Germany

Introduction: forced migration and civic engagement through food

In the aftermath of recent migration events, the reception and living conditions of asylum seekers and other migrants raise major political controversies and sociocultural concerns. In 2015, the number of people fleeing war, persecution, and lack of economic perspectives in the Middle East and West Africa increased substantially. Many migrants experience dangerous and traumatic journeys to reach Europe. Once in Europe, they face the challenge of inclusion and integration into society—from jobs, housing, and education to societal participation. Refugees are the target of infrastructures and policies aimed at monitoring, intercepting, immobilizing, and controlling migration (Bélanger & Silvey, 2020). These affect individual freedom of movement and residence, and put refugees in a situation of relative immobility amid a context of increasing xenophobia. Against nationalisms that demand the closing of frontiers and preaching the rejection of the Other, citizens in many countries are getting organized around projects that support the newcomers by introducing a "different normative vision of the future, in which notions of belonging and entitlement to rights are founded on criteria of residence, participation in community, and social relations developed in space and in relation to the 'commons'" (Nyers & Rygiel, 2012: 9).

Food occupies a prominent place in many initiatives for the support of migrants, which range from cooking workshops, community gardens, and food aid to refugee and ethnic food festivals (Food Relations, 2018; Veer, 2019). Both as a necessary form of sustenance and as a communicative system, food is considered by civil society organizations across the world to be capable of bringing people from different cultures and backgrounds together. The versatility of food and the diversity of its applications, be they noncommercial, job-related, or entrepreneurial, make the case for food to be thought of as corresponding to social integration. Food production, cooking, retail, catering, education, and shared commensality have, in this view, the potential to accomplish a sustainable role for the presents and futures of migrants in host societies. Nevertheless, the power of food for community cohesion and positive interethnic relations invoked by initiatives is often based on assumptions, ignoring the discrimination and othering that can occur through food (Wise, 2011). Still, as integration into a new society is an enormous challenge that necessitates a diversity of obstacles to be overcome, local citizens have pushed their creativity and engagement to enable active participation of the newcomers in the marketplace and in the public sphere, aiming to support sustainable presents and futures for migrants in host societies.

We see sustainability as a tripartite phenomenon with social, economic, and environmental dimensions (Purvis et al., 2019), where what is at stake is the capacity to continue a desired condition or process, to preserve a certain status, or to provide for the necessities of life (Tainter, 2006). To illustrate the interlocking of the social, economic, and environmental dimensions, consider the negotiations or conflicts between environmental advocates and rural people who live by natural resource production. These are "not just about ecology versus economics"; they are "also about sustaining a way of life" (Tainter, 2006: 92).

In this chapter we explore the role of food and mobile objects in the articulation of relations between migrants and host communities in the public space. Our focus is on two social integration initiatives taking place in Germany. Central to these projects are technological solutions that put food, human, and mobile objects together with the aim of overcoming some of the obstacles faced by newcomers. We are interested in how food, materiality, and movement operate and gain importance in the pursuance of contexts that favor sustainable livelihoods of newcomers, i.e., stable and safe lives of migrants in Germany.[1]

---

[1] This chapter is a modified version of an article published in 2022 under the title More-than-Human Assemblages and the Politics of (Food) Conviviality: Cooking, Eating, and Living Together in Germany, Food, Culture and Society (https://doi.org/10.1080/15528014.2022.2160893). This work has received funding from HERA JPR "Public Spaces: Culture and Integration in Europe" funded by the European Union's Horizon 2020 research and innovation program and the Deutsches Zentrum für Luft- und Raumfahrt e.V. under grant agreement number 769478.

The first initiative is RefuEat, a catering business in which the employees, all Syrian refugees, use bicycle kitchen trailers especially designed to travel on Berlin's roads and set up portable grill stalls at public events and private occasions. The second initiative is Über den Tellerrand (which translates as "[to look] over the edge of your plate," a German expression for open-mindedness), an association whose focus is to build community among people of different cultural backgrounds. With its sister organization, Kitchen on the Run, they use a shipping container with a built-in kitchen to travel through Germany and organize community-cooking events on their way.

Based on interviews with representatives of RefuEat and Über den Tellerrand and engagements in the field between 2019 and 2022, this chapter challenges the distinction between the materiality of the physical world and the social constructs of human intentions and accounts for how material objects can define relationships between different communities (Sautchuk, 2019). The conveyances devised by these two initiatives allow us to witness the entanglement of food, matter, and humans as they interrelate and bring about social configurations that respond to the unequal relations that affect the lives of refugees and migrants. In addition to New Materialist accounts (Govier & Steel, 2021; Kirksey & Helmreich, 2010), this study benefits from the insights of critical social movement studies, in particular those pertaining to the consequences of democratic empowerment and the exercise of citizenship that derive from "playing with" technology (Milan, 2016). More specifically, we explore the extent to which mobile materiality and food have an impact on migrant-local interactions in the public space and may thus be thought of as agentic objects that make things happen. The movement of the mobile kitchens of RefuEat and Über den Tellerrand through Berlin and across Europe makes us wonder: How does food, when bound up with other entities and issues, create connections and facilitate exchanges where previously there were none?

Although our approach recognizes the vital role of nonhumans, we want to avoid a framing in which shifts are emancipated and sustainable lives accredited mainly to material artifacts and technologies instead of the humans involved with them. Writing about the role of bicycles in early women's liberation movements, Furness (2005) alerts of the risk of overly accrediting female emancipation to the development of bicycles as a new technology. To avoid such a bias, he calls for situating discussions of the role of material agencies in social justice movements within their social and political contexts so that "it is possible to emphasize the emancipatory qualities of convivial technologies without resorting to a grand myth of technological liberation" (Furness, 2005: 410). Similarly, even though our focus lies predominantly on the recruitment of material artifacts and technologies in social initiatives toward minorities, we do not disregard the social, cultural, and historical factors, as well as the unequal power relations that shape these agentic objects and initiatives.

Pedaling food bikes to bypass immigration structures

RefuEat is a food business operating in Berlin since 2016. It was formed by Aymann, a second-generation German-Syrian designer, and Constantin, a retired German business consultant. In reaction to the lack of infrastructure to support the newcomers who arrived in Berlin during the long summer of migration, they wanted to create low-threshold employment opportunities for recent arrivals. Hence, they decided to found a business using customized kitchen bicycles to sell Syrian food all over the city, employing solely Syrian refugees. In 2019, RefuEat additionally opened an eatery and, more recently, due to the COVID-19 pandemic, developed a home delivery service. The dishes on their menu are common Middle Eastern street food, such as falafel and halloumi sandwiches, though much attention is given to details taken from traditional Syrian family recipes, for instance, adding pomegranate and lemon to their falafel sandwiches. Using bike trailers, RefuEat sets up portable grill stands at public events and private occasions. The bicycle trailers contain miniature kitchens, equipped with everything needed to prepare fresh food outdoors—from pans and deep fryers to washing bowls and a portable pavilion. As we will show here, these food bikes play an essential role in the world-making projects that happen through RefuEat.

For migrants and refugees newly arriving in Germany, building a life is hard—especially if their countries of origin lie in the Middle East, like Syria. Even after receiving refugee status in Germany, numerous hurdles exist that make it difficult for migrants to find employment. On an institutional level, these include, for instance, the lack of recognition of professional certificates and driver's licenses. Added to this is the deficiency in German language skills that makes it difficult for many to enter low-threshold jobs, and many migrants struggle to find employment. Well aware of these issues, Aymann and Constantin wanted to start a business that would bypass these obstacles and allow migrants to reach economic stability through their work, while simultaneously having the opportunity to improve their German skills and become part of a local community. Together, they developed the idea of building bicycle trailers that could move freely through the city and could be set up in different locations, requiring minimal investment and bureaucratic work. Riding bicycles does not require any form of license in Germany, and selling street food does not demand advanced German language proficiency. The food bikes allow RefuEat to circumvent many of the institutional obstacles refugee workers face. In that sense, they contain emancipatory qualities similar to those described by Furness (2005) in his study on bicycle-based activism and, more recently, by Rérat (2021) on the rise of the e-bike in urban space. By offering a more or less inexpensive alternative to modes of movements, such as the private car, which is a result of power relations, planning models, rules and social norms, and cultural meanings associated to it (freedom, social status, etc.), bicycles bring a drastic improvement in mobility to previously restricted communities. By affording increased freedom and emancipation to refugees, RefuEat's food bikes can be thought of as an emancipatory technology, shaping their immediate environment, fostering sociality, and helping to satisfy basic human needs.

Aided by a small electric motor that carries the weight of the heavy trailer, the RefuEat staff gains access to most of Berlin. The bikes are custom-made for the business, based on designs and ideas by Aymann and Constantin. In cooperation with a collaborative wood workshop in Berlin, they developed multiple models, each better tailored to the needs of the cyclists. This was a lengthy process because, although food bikes have been used by others in Germany before, they are relatively rare. Each enterprise also has food-specific requirements for the bicycles, such as space for a deep-frying pan in RefuEat's case, which makes the food bikes nonscalable. Similarly, local legal conditions, including the obligation to erect a shelter on-site in Berlin, make it necessary to customize the bicycles. Aymann, himself being a designer and believing in the importance of aesthetics, recruited a well-known tape artist to decorate the trailers. Inspired by street art styles, the striking, colorful patterns were specifically designed to attract the gaze of locals and meet the aesthetic ideals of an urban international clientele. As a social initiative, RefuEat also attracted the attention of the mainstream media and the state and civil society bodies, such as the Commission for Integration and Migration of the Senate of Berlin and the Expert Council on Integration and Migration. The civil society bodies endorsed the food bike catering activity, recognizing that language courses are not the only path to achieve social integration but that employment and participation in everyday life are also important aspects.

RefuEat's mobile menu is strongly shaped by the materiality of the food bikes. Due to the unavailability of extensive cooling facilities, they had to scrap meats from their menu, instead, they had to focus on vegetarian and vegan dishes like falafel and halloumi sandwiches. Despite (and because of) these restrictions, RefuEat has been able to create new and unique food networks through its integration of international ingredients and utensils. We have been told, for instance, about their tahini, which travels from Sudan to Lebanon before arriving in their Berlin kitchen; their dish packing boxes, made of palm leaf, compostable and fairly produced in India; or the hand-dried okra, which was brought back from Lebanon by a friend on holiday, who acquired it at a small local business. These materials become meaningful in that they put the far-away, for some even unreachable, on the map of Berlin. "[The okra] is really dried in our sun," Aymann told us one day. RefuEat thereby marks not only locations where one can consume Syrian dishes prepared by Syrian migrants but where one can participate in a whole international network of food links and processes.

By moving in brightly colored food bikes through the city, the refugees working for RefuEat are able to claim space and visibility outside of the usual service transactions. Utilizing movement in this way is a powerful tactic, if we think of movement as constituting a process through which social relations are made (or unmade), maintained, and performed (Urry, 2009). At the core of a movement is the human body, which "is not a stationary object but a lived sensuous subject in motion" (Christensen & Mikkelsen, 2013: 198), and through motion it creates places of meaning and opportunity, producing a sense of belonging (or not belonging) to a group, home, or community. The movement of food, ingredients, and recipes reflects the movement of the people involved with it. In this sense, the movement of refugees and food bikes through the city is not merely an economic strategy, but also a political comment: one of RefuEat's slogans *Streetfood aus der Heimat* (Street Food from the Homeland) reimagines public space by questioning what home means, who the streets are for, and how they can be experienced or practiced. By enabling these kinds of movements, the food bikes challenge the structures that organize and regulate the public space formally, which make business ownership and employment difficult for migrants. Simultaneously, through the catering activity, the workers also learn how to navigate and inhabit the private and public space within their new home. Sometimes this happens in ways that are not available to most migrants fleeing war and oppression, such as being present in art and design exhibitions, festivals, government premises, and brand and pop-up events. Other times this happens in ways that would not be possible even in their places of origin. By the end of July 2019, we accompanied the RefuEat staff to sell food during Christopher Street Day, the biggest LGBTQ + community event in Berlin. The massive gathering was an excellent occasion to generate high income. It also provided a context that people coming from a country with legislation that criminalizes the LGBTQ + community would have been less likely to experience in the public space of their hometowns. That day, we could observe a symbolic (dis)play aimed at representing social integration. One of the RefuEat owners bought one of the many rainbow flags available at the event and hung it on the stall in a visible place.

Exploring, experiencing, and working in the city is not a unilateral process, as the movement and trajectories of the kitchen trailers also change and rearrange foodscapes, which we understand here as the physical and social landscapes of food consumption and production. Through their mobile catering service, RefuEat adds a constantly changing spatial element to Berlin's foodscapes.

The map in Fig. 9.1 shows the locations where the RefuEat staff has been catering with food bikes from 2017 to 2020, as reflected on the Facebook page of the business. These places range from entertainment and leisure events, such as festivals on fashion design, contemporary art, and corporate events. RefuEat has also catered for events in venues of political power, such as several neighborhood town halls, the Press and Information Office of the Federal Government of Germany, or even the German Federal Chancellery.

The magenta marks in the map in Fig. 9.2 show the locations of asylum-seeker reception centers in the Berlin area where nonemployed refugees are expected to reside. These places are scattered at a distance

Figure 9.1 Berlin locations visited by RefuEat staff. *Based on data from Facebook and Google Maps.*

from the city center, in areas not only with less cultural diversity but also known to be less tolerant of or even hostile to refugees. The spatial dispersal of the reception centers makes the population of refugees fragmented and invisible (Fontanari & Ambrosini, 2018), a safer place away from the city center, which provides more opportunities for jobs and for general well-being.

The contrast is striking and well highlights the importance of mobility for achieving sustainable lives. The mobile materiality of the bike and the object's own affordances enable opportunities for refugees to exercise citizenship, which we conceive here as "a set of processual, performative, and everyday relations between spaces, objects, citizens, and noncitizens that ebbs and flows" (Spinney et al., 2015: 325). This way, the RefuEat staff is not merely integrated into the social fabric of Berlin but also has the possibility of becoming a part of its making.

Despite the challenge of operating self-made food technologies in a public space with strict regulations on street-food vending, RefuEat succeeded at constructing urgently needed niches within a capitalist system that is not intended to accommodate the needs of people in precarious circumstances, such as refugees. It similarly aims to claim space and social participation in a way that gives visibility to narratives that favor a view of social inclusion as an ongoing socio-material accomplishment constituted

Food on wheels: culinary paths toward sustainable lives for migrants in Germany 159

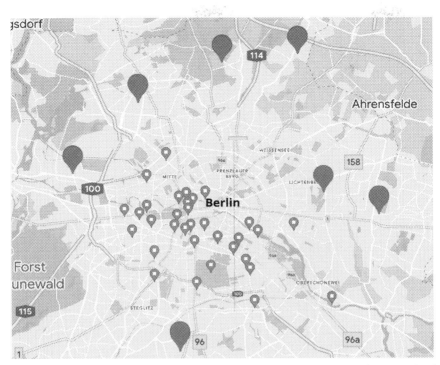

Figure 9.2 Asylum-seekers reception centers in Berlin. *Based on data from Facebook and Google Maps.*

and reconstituted as actors engaged in the world—a view in which horizontal living together prevails over assimilation. The food bikes are essential tools within these endeavors as they shortcut bureaucracy and provide a timesaver for people facing uncertainty regarding the length of their stay in Germany. The engagement with food and mobile technologies creates a pathway for refugees to claim space and social participation in the host society. It enables movement, visibility, and economic stability, which are preconditions for a more sustainable and socially just way of living.

Shaping food affects through mobility

Über den Tellerrand is a nongovernmental association established in 2014 in Berlin, which has more recently developed branches across Germany and internationally. Its aim is to create opportunities for people of different cultures, locals or migrants, to meet and get to know each other based on a principle of equality (https://ueberdentellerrand.org/start-englisch/). Food has played a prominent role since the organization's inception and development. It all started in 2013 when four Berlin-based students approached a refugee protest camp in Oranienplatz to support the claim for the rights of the refugees and establish neighborly relationships with them through cooking and sharing meals (Schöning & Hugelshofer, forthcoming). Convinced by the power of food to build bridges between peoples and cultures, the students continued and expanded the initiative under the legal status of an association, with food-related activities such as cooking classes, community cooking events, and urban gardening at its center. These events conform to what has been described as two conditions facilitating convivial encounters: built-in boundaries and shared purposes (Bredewold et al., 2020: 2057). The spaces defined by food and culinary practices (the kitchen, the dining room, or the urban garden), the clear purposes of the events, and, as shown below, the predefined roles assigned to participants afford encounters that are noncommittal: there is no pressure for the development of a deep, emotional connection but, instead, the togetherness experienced in convivial encounters is sufficient as an end in itself.

The cooking events allow attendees to meet in a context that spotlights the expertise and competencies of the migrants by platforming them as teachers. Migrants instruct attendants in cooking a recipe they themselves have decided upon. This enables a shift in narrative, moving away from the notion of locals and migrants as hosts and guests. "The idea is to break down the label a bit and not be either a refugee or a nonrefugee," explained Marieke Schöning from the Berlin headquarters (interview July 27, 2019). This way, migrants are able to leave their precarious, societally ascribed roles of guests at the mercy of the state's generosity. In the cooking events, the knowledge and skills of the migrants are recognized as valuable and meaningful by all attendants, who are there to witness them and learn from them. By bringing their own recipes and foods to the table, the "guests" themselves become "hosts" (Wise, 2011: 102) and interact on eye level with locals, who see migrants as equals rather than as the inadequate Other required to relearn everything in the hope of integrating through assimilation.

This backdrop that favors diversity and equality is the key to understanding Über den Tellerrand's actions not only as convivial and politically motivated, but also as an example of "radical cosmopolitanism" as it emphasizes "the desire to live and engage with others but also to be transformed by those considered as potentially different and as outsiders" (Baban & Rygiel, 2017: 101). Radical cosmopolitanism underlies the actions of socially engaged or "progressive" local/grassroots associations pursuing to bring people together through exchanges. Exchanges do not imply giving up hosts' and newcomers' own cultural particularities but a deliberate and organized effort to engage with these particularities. That is clear in Über den Tellerrand's culinary events as they push forward the "poetics of relation" (Glissant, 1997; Gutiérrez Rodríguez, 2015: 85) based on creative and affective crossings, and resist societal hierarchies of host and guest by sweeping away the binary divide between migrants and locals.

Kitchen on the Run is a portable version of the project, which develops within and around a shipping container with a built-in kitchen. Although their primary purpose serves the transportation of goods, containers have also become increasingly popular as portable infrastructures in the 21st century, serving, for instance, as temporary offices, accommodations, or classrooms. Their increased extra-industrial usage is what first raised the interest of Rabea Haß, Jule Schröder, and Andreas Reinhard, the creative team behind Kitchen on the Run during their planning phase. Inspired by shipping container architecture and design, the team wanted to create a project using a traveling container long before they had the idea to use it as a tool to advocate for the needs of migrants. Their focus on migration activism using food emerged out of the context of the 2015 migration events in Europe, from which the idea of the kitchen container as a place of encounter developed organically: "We were thinking we always invite friends to our kitchen, and it's really easy to get to know people while cooking together and sharing food" (Schröder, Interview: April 20, 2021). After successfully applying for funding and obtaining the donation for a container and kitchen utensils, the project became a reality: "the fat blue one" (*der dicke Blaue*, in German), a large, blue container arrived in Germany from China and joined in as the "fourth member team," in the words of Jule (ibid. interview). By perceiving the object as a teammate, the human members recognize its value beyond the pursuit of a shared goal and also as an opportunity to shape the team's affordances. The latter occurred in collaboration with additional individuals, namely, a class of architecture students at the Technical University of Berlin, to whom the human team members handed the nonhuman member over with a minimal brief: to design and build a portable, practical, and aesthetic space to cook, eat, and socialize in. Over a few months, the students integrated ideas into collaborative designs, resulting in a fully equipped kitchen with a dismountable wooden terrace and wooden folding furniture. Brainstorming sessions included cooking at Über den Tellerrand's kitchen in Berlin, with the aim of identifying any potential unease people might feel when cooking in an unfamiliar kitchen, such as opening cupboards outside of their own homes. To that effect, the designers developed an open-shelving system that allows one to easily see the contents of each storage unit without having to search around and open cupboard doors. The shelves were tailored to fit particular objects that would be traveling in the container, making tidying up at the end of events much quicker, and the transportation of utensils inside the container safer.

On its journeys, the kitchen container became a knot point for different movements of people, foods, and ideas, and facilitated the crossing of paths. Built for travel, shipping containers represent the quality of mobility, while simultaneously referring to the globalized economy. Despite being comparably simple in design in contrast with other traveling objects—a container has no engine, no wheels, no sails—profound changes in trade patterns and in the geographical location of economic centers over the last century indicate a strong connection between the emergence of the container and global trade (Levinson, 2016). The itinerant kitchen container, too, is highly symbolic, as it underlines the strong contrast between the immobility of the refugees and the expansive geographies of global supply chains. In 2016, Kitchen on the Run traveled through Europe, visiting the cities of Bari in Italy, Marseille in France, Duisburg in Germany, Deventer in the Netherlands, and Gothenburg in Sweden (Fig. 9.3). The international journey followed a route from Southern to Northern Europe with the specific purpose of tracing one of the most used routes by fleeing people and the European countries that, at the time, received the largest number of migrants. The route was intended to raise the question, "If goods can cross borders, then why not people?" (Schöning, Interview July 27, 2019). During the European tour, the team stayed in each location for four weeks and organized community-cooking events that are similar, both in objectives and approach, to those held in Berlin and by satellite organizations: The container was the site of regular, free cooking events at which migrants were invited to teach a recipe from their country of origin to the other participants. As told by a Kitchen on the Run organizer, this tour was the first opportunity for some refugees to prepare "their" food in several months, since not all refugee camps or shelters provide the possibility of cooking. The event organizers even invited the refugees to cook especially complicated and time-consuming recipes so as to keep all participants busy and happily engaged in all activities, from the *mise en place* to eating to dishwashing and cleaning.

Figure 9.3 Kitchen on the Run 2016 Europe Tour. *Courtesy from Über den Tellerrand/Kitchen on the Run/Amelie Persson.*

Since 2017, the team has been touring four German locations each year (with the exception of 2020 and 2021), most often in small towns and cities, because of their more limited exposure to cultural diversity and because the participants were more likely to meet again. The main goal in Germany was to create encounters between locals and people with a migrant background and to foster community building. In the best-case scenario, the container would leave behind a "cooking community" that continues to gather and that might even start a new satellite of the Berlin-based association. Besides migrants and refugees, the cooking events were typically attended by locals from "the classical engaged social strata," as one of the organizers put it, by people who were already familiar with social justice issues surrounding the livelihoods of migrants in Germany, and who were invested in community activism. This stresses the political dimension within these events, reminding us that not everyone has the same opportunities to move, interact, and construct a context for being with, caring for, and receiving care from others.

Even though many event attendees are already familiar with the social justice issues that motivate the container's travels, its integration into public foodscapes enables its reach to far exceed this group of people. Public foodscapes foster encounters that do not require intentionality; smells and other sensory traces of food have the power to spontaneously draw people into places and relations they did not intend to seek out (Wise, 2011). The unusual design and aesthetics of the kitchen container enhance this power to draw the attention of passersby and to recruit them into their project. This potential is furthered as the container merges elements of private and public foodscapes by bringing a representation of the domestic sphere—a homely kitchen space—into a public place. The cooking classes, in addition, allow foods and food practices to transition from one realm into the other, as previously alien or unapproachable dishes and ingredients become familiar and attainable. By participating in the preparation of such foods, their parts in local foodscapes—the shops where their ingredients can be bought, the kitchens that sell these dishes, the people that prepare them, etc.—are no longer representations of the Other but instead offer possibilities for engagement and participation. Consequentially, the foodscapes themselves become deanonymized as the people, foodways, and food stories they are constituted with, gain in visibility and connectivity. The itinerant kitchen achieves this by creating a point of contact between persons whose lives might rarely cross otherwise. And it does even more than that: The kitchen container functions as a kind of agora, in the sense of an open place for assembly. Indeed, although it is limited by its perimeter, the mobile kitchen does not have a rigid function as a place for cooking and eating. Besides the culinary events, Kitchen on the Run uses the container for activities aimed at fostering exchanges that arise through encounters, such as the "human library," a tool that seeks to challenge prejudices and stereotypes, consisting of volunteers and participants narrating stories and learning from each other. By doing so, Kitchen on the Run pushes the social boundaries and meanings of commensality, the practice of eating together, to potentially become a generator of new sociality and participation in the public space.

Through its movement and its aesthetic and practical affordances, the container passes through communities; encodes biographies, memories, and identities; cumulates positive affective values; and becomes entangled with change (Given, 2013). This positive affective value does not exist in isolation but emerges through interaction and collaboration between human and nonhuman agents (Kipnis, 2015). And although the results of collaboration may not always be positive—interactions can be dissonant, discordant, or disruptive—the Kitchen on the Run team did not witness

Food on wheels: culinary paths toward sustainable lives for migrants in Germany   165

any hostility. There were plenty of emotionally demanding experiences, but none that were antagonistic.

In addition to its function as a traveling kitchen, the container has fulfilled other purposes more recently. During the summer of 2020, it was featured as part of an architecture and design exhibition at the Pinakothek der Moderne, the museum for modern art in Munich. Its placement at one of the world's largest museums for modern art offers a perspective that sees the container itself as a work of art. In a way, this closes the circle to the origin of the project: Since its beginning, the container has been an inspiring object whose affordances not only include the practical purpose of transport but simultaneously have creatively stimulating affects. Like a canvas, it has absorbed layers upon layers of stories and meanings on its travels. Alluding to the ways in which people are denied rights and opportunities, the container stimulates the beholders' creative perception, inspiring them to "imagine a world with room for everyone and everything" (Ingold, 2020: 604). The container being featured in an exhibition at the Pinakothek der Moderne highlights its ability—physically and affectively—to move and to be moved.

Conclusion

The activities of RefuEat and Kitchen on the Run toward newcomers rely heavily on the affordances of the mobile objects they work with. RefuEat's food bikes are precious tools for the migrant workers to navigate the city and Germany's bureaucratic system: They provide them with sustainable lives by facilitating mobility within the city, economic stability, and social connectivity. Through these dimensions they also increase the possibilities of agentive actions of the workers, who, through their movement, become able to participate in the performance of Berlin's public space. Kitchen on the Run's traveling container pursues a similar objective by offering a platform where people from different cultures can get together, and the migrants can enact agency.

Both initiatives provide economic sustainability and opportunities for encounter and community building through contexts and practices that enable migrants to live on their own terms, rather than in dependence on a generous host society. By reducing economic dependency on institutional support, they help migrants to regain autonomy over their lives. Bringing together migrants and locals on equal terms and in a way that highlights the expertise of the migrants, they furthermore work to deemphasize the notion of assimilation and instead give priority to the cocreation of social inclusion.

We appreciate that a focus on mobile materiality does not necessarily offer full-scale solutions to the many challenges of inclusion refugees and migrants face. Nonetheless, our study sheds light on the ways in which people from different communities are trying to collectively solve those issues by promoting encounters and interactions through the medium of food and movement. The two initiatives described here suggest that the assemblage of people, food, and mobile objects can strongly influence the relations between migrants and locals by creating opportunities to circumvent the social, spatial, economic, and bureaucratic hurdles imposed on the former. RefuEat and Kitchen on the Run have constituted mobile nodes and practices within foodscapes that encourage relational and affective experiences, which, in turn, bring about new possibilities of being in the world.

References

Baban, F., & Rygiel, K. (2017). Living with others: fostering radical cosmopolitanism through citizenship politics in Berlin. *Ethics & Global Politics*, *10*(1), 98–116. Available from https://doi.org/10.1080/16544951.2017.1391650.

Bélanger, J., & Silvey, R. (2020). An Im/mobility turn: power geometries of care and migration. *Journal of Ethnic and Migration Studies*, *46*(16), 3423–3440.

Bredewold, F., Haarsma, A., Tonkens, E., & Jager, M. (2020). Convivial encounters: Conditions for the urban social inclusion of people with intellectual and psychiatric disabilities. *Urban Studies*, *57*(10), 2047–2063.

Christensen, P., & Mikkelsen, M. R. (2013). 'There is nothing here for us.!' How girls create meaningful places of their own through movement. *Children & Society*, *27*(3), 197–207.

Fontanari, E., & Ambrosini, M. (2018). Into the interstices: Everyday practices of refugees and their supporters in Europe's Migration 'Crisis'. *Sociology*, *52*(3), 587–603.

Food Relations. (2018). Food as a key medium for social inclusion and intercultural dialogue: A taste of lessons learned from European initiatives. Freiburg: Agronauten.

Furness, Z. (2005). Biketivism and technology: Historical reflections and appropriations. *Social Epistemology*, *19*(4), 401–417. Available from https://doi.org/10.1080/02691720500145696.

Given, M. (2013). Commotion, collaboration, conviviality: Mediterranean survey and the interpretation of landscape. *Journal of Mediterranean Archaeology*, *26*(1), 3–26. Available from https://doi.org/10.1558/jmea.v26i1.3.

Glissant, É. (1997). Poetics of Relation. Ann Arbor: University of Michigan Press.

Govier, E., & Steel, L. (2021). Beyond the 'thingification' of worlds: Archaeology and the new materialisms. *Journal of Material Culture*, *26*(3), 298–317.

Gutiérrez Rodríguez, E. (2015). *Archipelago Europe: On creolizing conviviality*. *Creolizing Europe. Legacies and transformations, edited by Encarnación Gutiérrez Rodríguez and Shirley Anne Tate* (pp. 80—99). Liverpool: Liverpool University Press.

Ingold, T. (2020). Art and anthropology for a sustainable world. In V. Oliveira Jorge (Ed.), *Modos de Fazer/ways of doing* (pp. 603—622). Porto: CITCEM.

Kipnis, A. (2015). Agency between humanism and posthumanism: Latour and his opponents. *Hau: Journal of Ethnographic Theory, 5*(2), 43—58. Available from https://doi.org/10.14318/hau5.2.004.

Kirksey, E., & Helmreich, S. (2010). The emergence of multispecies ethnography. *Cultural Anthropology, 25*(4), 545—576.

Levinson, M. (2016). *The box*. Princeton: Princeton University Press.

Milan, S. (2016). Liberated technology: Inside emancipatory communication activism. In E. Gordon, & P. Mihailidis (Eds.), *Civic media. Technology, design, practice* (pp. 107—124). MIT Press.

Nyers, P., & Rygiel, K. (2012). Introduction. In P. Nyers, & K. Rygiel (Eds.), *Citizenship, migrant activism and the politics of movement* (pp. 1—19). London: Routledge.

Purvis, B., Mao, Y., & Robinson, D. (2019). Three pillars of sustainability: in search of conceptual origins. *Sustainability Science, 14*, 681—695.

Rérat, P. (2021). The rise of the e-bike: Towards an extension of the practice of cycling? Mobilities. Available from https://doi.org/10.1080/17450101.2021.1897236.

Sautchuk, C. E. (2019). The pirarucu net: Artefact, animism and the technical object. *Journal of Material Culture, 24*(2), 176—193.

Schöning, M., & Hugelshofer, V. (forthcoming). Encounters. In R. Dolphijn & N. Polson (Eds.), *Negotiating foodscapes: A glossary*. Cham: Palgrave MacMillan.

Spinney, J., Aldred, R., & Brown, K. (2015). Geographies of citizenship and everyday (im)mobility. *Geoforum; Journal of Physical, Human, and Regional Geosciences, 64*, 325—332.

Tainter, J. (2006). Social complexity and sustainability. *Ecological Complexity, 3*, 91—103.

Urry, J. (2009). Mobilities and social theory. In B. Turner (Ed.), *The New Blackwell companion to social theory* (pp. 477—495). Sussex: Blackwell.

Veer, E. (2019). Fostering community values through meal sharing with strangers. *Sustainability, 11*, 2121. Available from https://doi.org/10.3390/su11072121.

Wise, A. (2011). Moving food: Gustatory commensality and disjuncture in everyday multiculturalism. *New Formations, 74*(1), 82—107.

CHAPTER TEN

Agri-food routes as tools for sustainable rural development: the case of chili route in Yahualica Denomination of Origin

Daniel De Jesús-Contreras[1], Laura Elena Martínez-Salvador[2], Emerio Ruvalcaba-Gómez[3] and Frédéric Duhart[4]

[1]Centro Universitario Temascaltepec. Universidad Autónoma del Estado de México, Toluca, Mexico
[2]Instituto de Investigaciones Sociales. Universidad Nacional Autónoma de México, Ciudad de México, Mexico
[3]Centro Agroalimentario CAXCAN, Jalisco, Mexico
[4]Sigmund Freud University, París, Francia

Introduction

In 2018, the Mexican Institute of Industrial Property (IMPI) granted the Denomination of Origin (DO) to the Yahualica chili (*Capsicum annuum*), which is produced within the limits of a territory comprised of nine municipalities from Jalisco (Los Altos Region) and two from Zacatecas. Until then, only one type of chili, the Habanero (*Capsicum chinense*) from the Yucatan Peninsula, had this institutional protection granted in 2010. This is of particular interest since Mexico is one of the main centers of origin and diversification of *Capsicum annuum* (Aguilar et al., 2018). So it is striking that only one species has this distinction compared with the nineteen varieties of chili that currently benefit from the Protected Designation of Origin or a Protected Indication of Origin in the European Union.

In the protected area under the Yahualica DO, it seeks to become an important tool to value primary production, stimulate the productive interrelationships of the territory, promote value addition, improve commercialization, and protect the product from unfair competition. Parallel to the DO, for years, various actors in the protected territory have undertaken actions aimed at protecting, safeguarding, and promoting the agri-food goods. Within these

---

Food, Gastronomy, Sustainability, and Social and Cultural Development.
DOI: https://doi.org/10.1016/B978-0-323-95993-3.00003-7

© 2023 Elsevier Inc.
All rights reserved.

169

strategies is the touristic, territorial, and heritage activation of Yahualica chili, through which it seeks to influence the strengthening of the productive chain and promote a holistic concept of rural development from different perspectives — agri-food activity, conservation and restoration, and cultural heritage.

Thus, the agri-food routes, especially the Chili Route, emerge as tools of support, visibility, and valorization of chili itself and of the figures of the DO. In the context of the emergence of new socioeconomic functions of the territory, the valorization of the intangible heritage of rural areas is seen as one of the main development alternatives (Delgado & Hernández, 2019). Agri-food products with territorial anchorage and identity roots are positioned today as important development assets for rural sectors.

This paper deals with a qualitative and exploratory case study (Yin, 2003) focused on reflecting, from a sociocultural perspective of sustainability, on the role of agri-food routes in the rural development of territories that host distinctive agri-food products (De Jesús-Contreras & Medina, 2022), specifically the Ruta del Chile in Yahualica DO, located in Mexico.

In this regard, it is important to mention that this research has been built from a multidisciplinary perspective, resulting from an open and constant dialog between a team of researchers from disciplines such as anthropology, economics, and rural sciences and a key actor who assumes the role of incidence in the processes of territorial transformation. Therefore the research is also based on an action-research exercise, since it implies a self-reflective and critical process to understand the reality that affects the territory (Melero, 2012).

Agri-food routes and sustainable rural development

Since the mid-1990s, it has been confirmed that the classic model of rural development, reflected in the Green Revolution, has been insufficient in improving the conditions of marginalization and poverty in rural societies. It has even accentuated social asymmetries in access to the means of production and triggered other harmful effects on health and the environment (Herrera, 2013). Faced with these problems, almost in parallel, new paradigms of rural development began to permeate academic and nonacademic discussion spaces.

These new perspectives propose the idea of a rurality away from the simplistic agricultural approach and support the premise of the multiplicity of functions that the countryside acquires, as well as the insertion of rurality in more complex markets. Some concepts, such as new rurality (Carton de Grammont, H, 2004), new agrosocial paradigm (Monllor, 2013), rural territorial development (Schejtman & Berdegué, 2004), or quality turn (Goodman, 2003), were proposed to show the structural changes of rurality and the need for new analytical frameworks for its interpretation.

One of the priority topics in these new approaches is the sustainability of agri-food systems and territories to guarantee, in the medium and long term, food sovereignty and the continuity of agricultural practices as a means of territorial reappropriation. Sustainable rural development raises the idea of a "new rural culture" and a scheme of human, economic, and social development capable of continuing, under the concept of harmony, with the ecological, cultural, and productive systems of the rural territories. Although primary activities continue to be the dominant axis, the new rural culture recognizes the existence of other forms of production of added value in rural spaces — closely linked to the concept of territoriality — and implies the construction of capacities and the revitalization of the peasant economy (Zambrano et al., 2015).

Therefore, it could be understood as a social transformation linked to economic growth that seeks a balance between the use of resources (culture, landscape, natural heritage) and the needs of rural communities to mitigate the problems associated with poverty, marginalization, and lack of opportunity. From our point of view, sustainable rural development makes sense if: (a) agriculture is articulated with other sectors to guarantee the continuity of the productive structure; (b) innovative forms of development linked to a territorial vision are promoted; and (c) true inclusive projects are generated that involve strengthening social capital and building capacities (Zambrano et al., 2015).

Within the new strategies for rural development, we find initiatives associated with the activation and sociocultural valorization of food heritage, especially since rural territories are home to agri-food products capable of boosting the local economy and encouraging collective action for their valorization. We can see how, for a few decades, different public, and private actors, related to rural development, the agri-food sector, and even tourism, have been adding value to their heritage through quality labels (geographical identifications, trade seals fair), rural tourism (interpretation centers, museums, festivals), among other heritagization actions (inventories, recipe books).

Agri-food routes can be understood as an itinerary organized around a product or a group of products representative of the culture and identity of a territory. These are made up of producers, agroindustrial companies, regional restaurants, and other service providers (Murgado et al., 2012). Its objective is to boost regional economies through the valorization of food and local gastronomic culture, stimulate the development of food quality labels and consolidate the productive culture of the territory (Millán et al., 2011). There are routes by product (wines, cheeses, tequila), by dish (*Ruta del Mole, Ruta del Cassoulet*), and by ethnic-gastronomic content (*Ruta de los Mayas y el chocolate*), although there are also routes by thematic content (route of the markets, route of the convents) (Tresserras & Medina, 2007).

It is also important to highlight that the agri-food routes that operate in rural and urban spaces present different levels of institutionalization, since some are formalized and have financial support or are backed by promotional organizations. Whereas others operate informally or, in some cases, have not materialized (Suremain de, 2017), and function as support elements to position the discourses of political actors or business groups.

Agri-food routes are potential tools for sustainable rural development because they value local agri-food production, promote the articulation of different sectors, promote cooperation between actors, and contribute in boosting the local economy. The food routes can be useful to form a structure of actors organized through formal and informal agreements, which can gradually translate into a group from which it is managed. This means that these routes can benefit the entire territory through the enhancement of other products and services and the integration of other economic sectors (Brunori & Rossi, 2000).

Denomination of Origin Chile Yahualica

The Yahualica chili belongs to the species Capsicum annuum L., a variety classified as neotropical (Aguilar et al., 2018). The production of the chili involves mass selection, a technique practiced by peasant communities to choose plants and seeds with better character and agronomic potential. However, in terms of genetic diversity, it is highly probable that the chili species cultivated in Mexico are a product of a constant and historical genetic erosion (Rodríguez, 2018).

The Yahualica chili makes use of artisanal techniques for its cultivation and harvest, such as drying in the sun and laying out on blankets, so it maintains its shape and texture for longer (Miranda, 2020). These elements have earned it institutional protection under the figure of the DO, granted by the Mexican Institute of Industrial Property since March 16, 2018 (Secretaría de Economía SE, 2018).

According to the Regional Center for Integral Services for Protected Agriculture (CRESIAP) (2020), there are around 64 chili-producing communities, concentrated, with greater density, in the municipalities of Yahualica, Teocaltiche, Cuquío, and Nochistlán. Chili production is concentrated in numerous spaces in the region close to bodies of water and located transversely in a kind of productive corridor. However, the evolution of chili production in the municipalities that make up the territory of the DO has shown an erratic trend. It is important to mention that this information considers only the variety of chili that could be protected under the DO Yahualica, and that in the official statistical bases it only includes dried chili.

After 2017, the municipality of Yahualica presented a considerable increase in chili production, going from 365.2 tons in 2017 to 1135.9 tons in 2018, with a new decrease in 2019. The second place in the production of chili is occupied by the municipality of Teocaltiche, and the third by Encarnación de Díaz (Secretaría de Agricultura y Desarrollo Rural SADER, 2020). This increase in production and subsequent decrease in the municipality with the highest production was mainly due to the inclusion of chili in the public agendas, as it was made visible with the DO. However, a series of problems and challenges, mainly caused by disarticulation with other links in the agri-food value chain, especially with food markets, caused an overproduction of chili, which ruled out subsequent planting and production.

The foregoing discussion, within the sustainability framework, calls into question the need to build a value chain that transitions to an agri-food production chain with a territorial approach.

Among the multiple challenges the local production of Yahualica chili faces is the atomization of the cultivation plots, which implies a considerable universe of small producers. Young generations do not show interest in productive activities related to chili, and current producers show reluctance to technical and technological changes in the links of production and transformation (Centro Regional de Servicios Integrales para la Agricultura Protegida, CRESIAP, 2020). The producers are aware of the common problems (lack of agricultural labor, migration, market losses to foreign products), but there is still a great dispersion of efforts and "little organization" (Centro Regional de Servicios Integrales para la Agricultura Protegida, CRESIAP, 2020: 90).

Chili production faces a socio-environmental problem, particularly due to the constant use of highly toxic and potentially carcinogenic fertilizers and herbicides (BBC News Redacción, 2020). This is significant because more than 59% of the soils in the Altos del Norte Region and 87% of the Altos del Sur present degradation processes, most of them associated with the use of chemical fertilizers (Centro Regional de Servicios Integrales para la Agricultura Protegida, CRESIAP, 2020: 53). However, due to the land tenure regime, there is a crop rotation system, which could allow the soil to have periods of natural buffering of contaminants.

There is shortage of water in the Altos Sur area of Jalisco because the region is a semidesert area, a border region between Mesoamerica and Aridoamerica. The region faces increasing problems of contamination due to the organic wastes and nitrates resulting from the incorrect disposal of wastewater and runoffs from agricultural areas Secretaria de Medio Ambiente y Recursos Naturales (SEMARNAT), REq Consultores, Gobierno del Estado de Jalisco (GJAL).

The persistence of these socio-environmental problems shows that the DO can be an effective tool for the sustainability of the territory and the agri-food system of chili. However, it is necessary to point out that, after the declaration, other problems have arisen. The DO is not meant to effectively integrate all the actors involved in the production chain, since from the beginning it was limited only to the primary sector (chile producers). It avoids the field of agribusiness where various salseros are integrated, contributing to building a reputation. In fact, a significant proportion of the foodscape around chili is due to its commercialization in small containers of sauce and ground or roasted chili, which indicate the origin.

Finally, the DO has sparked disputes between different power groups over the appropriation, administration, and management of the food quality label. A self-appointed group, the Regulatory Council of Chile Yahualica, has maneuvered early and opportunistically, taking advantage of national legal shortcomings, which is inconsistent with international protocols for the formation of a true regulatory council, and tries to claim the almost exclusive use or normative of the DO.

The Secretary of Agriculture and Rural Development (SADER) Jalisco, for instance, warns that a body such as the Regulatory Council cannot be integrated since the Mexican Standard, which will legislate the conditions and requirements of the DO, has yet to be published. A true regulatory council of Chile Yahualica would be the official body in charge of watching over the producers' interests in adhering to the DO, supervising its proper use, and sanctioning those who make improper use of institutional protection.

This scenario of ecological problems, exclusions, and social disputes poses serious challenges to the sustainability of the DO and questions its functionality in rural development processes.

Gastronomic and food uses of Yahualica chili

For Mintz (2003), what a society produces, and eats can be considered a substantial part of its culture. So it is possible to affirm that in the territory of Yahualica, a true culture of chili consumption has been configured because the product is linked to different spheres of daily life (agricultural production, economic activities) in the territory. However, its role in food culture is particularly important. In relation to tourism, food plays a determining role. With more and more tourists devoting a significant percentage of their spending on food, it is even more interesting that some segments express that the main motivation of their trip is to know and taste a particular food (Medina, 2017).

In the case of Yahualica chili, artisanal processes predominate in the selection of seeds, with great participation of women and children (Centro Regional de Servicios Integrales para la Agricultura Protegida, CRESIAP, 2020) for harvesting and picking by hand. The artisanal production processes of chili affect its differentiated quality because they reflect the territorial anchorage and the transmission of know-how, positioning it as an important local heritage.

It is important to mention some elements of the Yahualica chili in its gastronomic aspect. Despite its spicy properties, this chili does not have the adverse physiological effects of some other varieties, such as tachycardia, extreme sweating, or stomachaches (García, 2017). Another of its peculiarities, in addition to its flavor, is its characteristic morphological consistency, which differentiates it from other similar products on the market. The Yahualica chili is the basis of some typical dishes of Jalisco cuisine, which has contributed to the tourist positioning of this Mexican state. An example of this is the well-known *tortas ahogadas*, one of the most deeply rooted foods in the state, especially in Guadalajara, which consists of salty bolillos stuffed with carnitas and cured onions and submerged in a sauce whose main ingredient is Yahualica chili. This is one of its main differentiators in terms of flavor and hotness. The chili is used as a base to make the birria, the Jalisco pozole, the *mole de olla*, and the *menudo*, among other traditional dishes.

Among the most popular forms of consumption are different sauces that vary in presentation, heat, and ingredients. Indeed, the Yahualica chili has fostered the development of rural agroindustries oriented to the processing and preparation of sauces that are sold in presentations of 355 milliliters to one liter. A tour of the Yahualica municipal market allows you to observe with obvious clarity the diversity of sauces and brands made by local entrepreneurs.

Similarly, in the gastronomy of the region, sauces are used to season or garnish other foods, and it is worth mentioning that currently a process of gourmetization of chili is also observed, partly driven by the denomination of origin effect. Some local entrepreneurs have diversified their consumption with products such as ground and roasted chili, macha sauce, or chili salts, sold outside the region and in virtual stores specializing in gourmet products.

The presence of chili in the local cuisine suggests its continuity within the cultural and productive structure of the territory, but at the same time, it is valued by other segments of consumers, who could raise new views on a local product that is inserted in other sociocultural and economic dynamics of a global nature (Appadurai, 2001).

The insertion of food goods into the tourism market is a sample of the circulation of imaginaries and cultural symbols that are demanded for their authenticity (Lash & Urry, 1994). This imposes some challenges for the sustainable development of rural territories that suffer the effects of their insertion in globalization through the DO system and tourism.

Tourist and gastronomic activations of Yahualica chili

In the context of the economic and cultural appreciation of the Yahualica chili from the DO, its territorial activation as a tourist attraction is taking a significant rise. It is not surprising that obtaining this quality label results in a central strategy to trigger other alternative projects that contribute to the positioning of the product and the territory, as it happens quite frequently in other cases of DO in Europe. (Millán et al., 2011).

The DO of the Yahualica chili has motivated other actors in the territory, linked to the gastronomy sector, tourist services, and even cultural heritage, to have undertaken valorization strategies that position the chili beyond food or as an ingredient in the local cuisine, elevating it to the category of tourist and gastronomic attraction. This does not mean that prior to the denomination of origin, chili was not relevant within the tourist-gastronomic offer of the territory. However, this institutional protection device implied a phenomenon of positioning an agri-food product that until then seemed to stick to its territory of origin.

It is important to note that since 2015 (except for the 2020 edition) the *Fiesta de Todos los Chiles* has been celebrated in the municipality of Yahualica, an event to promote not only the product, but all the gastronomy of the area. In this event, the historical and family links of producers with chili are disseminated, and tradition and family heritage are highlighted, elements that play a fundamental role in the formation of human resources responsible for perpetuating cultivation practices and food uses of the product (García, 2017). Likewise, this event takes advantage of the great tradition of local patron saint festivities as a precedent so that, parallel to the DO, a Yahualica Chili Route is implemented. This route is conceptualized as a rural and gastronomic tourism project focused on the culture and productive chain of chili.

It is necessary to indicate that, although this agri-food route is currently in an experimental and settlement phase, its implementation is part of other religious tourist routes, very traditional and with a high degree of consolidation. The region of Los Altos de Jalisco is recognized for being part of the cradle of the Cristero War, which promoted, for over seven decades, the recognition of spaces as cultural heritage, having been valued through tourist itineraries throughout the most representative sites of the region. In this way, some municipalities included in the DO of the Yahualica chili converge with the area of influence of these religious routes and represent a continuous flow of tourists.

The tourist activation of the Yahualica chili is inserted into a pre-existing tourist market and at the same time contributes to diversifying the tourist offer of the region. However, the route acquires its own personality and a particular form of management because its intermediaries are more closely related to the agri-food, agroindustrial and gastronomic sectors.

The Yahualica chili route and sustainable rural development

The construction of agri-food routes around a representative product of a certain territory refers at the same time to an example of the insertion of rural spaces into globalization (Suremain de, 2017). This means that rural development processes are a materialization of globalization in local contexts (Appadurai, 2001). This perspective assumes that the sustainability of the DO and the agri-food routes must necessarily be situated in a constant dialog between the global-local dimension instead of assuming a dissociation between them.

Thus, sustainable rural development can be achieved if there is a link between the agricultural sector and other economic sectors that makes it possible to boost the productive structure through the promotion of innovative forms of development linked to a territorial vision and materializing in inclusive projects. In this sense, the chili route can attend to these elements in the following way:

1. Linking the agricultural sector with tourism, gastronomy, and heritage. The Yahualica chili route is a project articulated around an agri-food product whose quality and sensory specificity are linked to the territory and to a particular know-how. This route extends through ten of the eleven municipalities considered in the territory of the DO, so it is a route that acquires a wide geographical dimension and that, for technical purposes, can be ambiguous and imprecise for the experience anchored to a specific tourist space. A significant part of the experience on the route includes tours of the cultivation fields, where chili production is located, to observe the artisanal processes and appreciate the agri-food landscapes. In this sense, knowledge of the territorial origin of chili has an important impact in terms of the sustainability of the agroecosystem where it is produced. In addition, recreational activities contribute to the conservation of natural resources strengthening the social, institutional, and productive social relations, as well as the continuity of cultural practices.

   The realization of the *Fiesta de Todos los Chiles*, and a gradual implementation of the route, influenced the actors linked to the restaurant and gastronomy sector to undertake a movement of valorization of chili within the local cuisine. For example, the *birrierias* and *menuderias* located in the municipality of Yahualica today make it explicit that their dishes are prepared in the traditional way with authentic Yahualica chili. In the same way, several *kioskos* have been installed in the portals of the municipal market in which chili is exhibited, in addition to the fact that the producers of the sauce containers highlight the links of the product with the territory.

It is possible to perceive that these communication strategies of territorial origin are directed more toward tourists than residents, for whom the presence of chili in traditional foods is inherent. Gastronomy linked to tourism acts as a territorial identity marker for chili. In a broader sense, the route works as a tool to territorialize the product in its place of origin and favor the retention of added value and the articulation of the agricultural sector with the gastronomy sector. Synergies are created in this route in the form of short marketing chains, which promote the producer-consumer relationship, reducing the presence of intermediaries to promote more dynamic alternatives for activating territorial resources. This has an important resonance for sustainable rural development because it strengthens a new rural culture through the diversification of the functions of agriculture. The irruption of rurality away from a sectoral and developmentalist vision solely focused on agricultural production.

2. Territorial innovation processes. Although the route under construction already constitutes a sample of the innovation processes, it is important to consider other manifestations that also show the capacity of the actors to generate socioeconomic diversification strategies. The route is generating a territorial valorization and innovation scheme based on the construction of a basket of goods and services (Pensado & Martínez, 2015). Other agri-food products of the region, such as milk candies, sweet potatoes, pumpkin, cheeses, peanuts, *tacazotas*, sausages, jamoncillos, cajeta, pork rinds, black corn pozole, and Tequila itself, find a significant market, complementing the tourist and the gastronomic offer associated with chili.

The valorization of chili, within the framework of the DO and tourism, generates new gastronomic uses. Its incipient incorporation into the gourmet product market opens the possibility of its integration into *haute cuisine*, thereby creating other alternatives for valorizing a local product. Chili goes from being just an agricultural and food resource limited to a very local environment to positioning itself as a cultural and tourist consumer goods, with circulation in a global market hungry for authenticity. For example, some products made with chili sauce containers — are transformed into souvenirs in response to the demands of a market that seeks to prolong the gastronomic experience through souvenirs, but at the same time acquires the role of tourism promoter of the territory (Espeitx, 2004).

Of course, these forms of territorial innovation are generating effects on rural development because they promote new value-added strategies, representing sources of employment and areas of opportunity for segments of the population, such as young people or returned migrants.

3. Inclusion of territorial actors. One of the weakest points of the route refers to the unequal integration of the actors involved in the different links of the productive chain. In the first place, it should be mentioned that this project arises from local governments as a partial response to the problems expressed by chili producers related to the absence of stable marketing schemes, lack of functional organizational structures, and a weak economic link with the markets. The route takes a clearly vertical orientation with the intention that the producers take ownership of the project and adapt it to their individual needs, rather than a collective project built participatively from the identification of common problems.

Second, the route is operating on a significant social fracture, arising from a series of disputes between actors for the award of the DO. This situation is aggravated by the existing asymmetries and builds a field of constant conflicts for the appropriation of the territorial reputation of chili now associated with the route. This situation of exclusion is not new. Since the formulation of the regulatory framework of the DO, the group of salseros has been left behind, a situation that currently prevails.

Likewise, there is a latent risk of deterritorialization of the Yahualica chili, especially if the regulations ignore elements of the production chain, such as allowing the packing or packaging stages to take place outside the area protected by the DO; a regulatory gap that may be of special interest to foreign business people. However, in the case of the route, it is the actors from the agroindustrial and restaurant sectors who are capitalizing on the benefits of the DO from an alternative position linked to tourism, an activity that has provided a window of opportunity for the excluded actors.

Finally, the processes of territorial innovation mentioned suggest the emergence of new actors in the territory, as well as the need for a strong articulation between the different agents, including the public and governmental institutions of the territories in the DO, who need to generate mechanisms of equitable appropriation of a heritage that has been built collectively. The gastronomic and agroindustrial activities on the rise around chili acquire an exclusive character, leaving a marginal benefit for the producers. In this case, there is a risk of having a chili route without the effective presence of primary producers, as observed by Thomé et al. (2014) in the case of the nopal route in Mexico City.

Conclusions

In this paper we analyze the case of the Yahualica chili route to understand how the link between food heritage and tourism can affect the development of rural territories. We also highlight the challenges to the sustainability of agri-food systems. The sustainability of the territories, especially those where the trinomial food-culture-economy, is interwoven to create tourism-agri-food dynamics is complex due to all the social actors, the productive dynamics, and the institutional frameworks that impact, promote, and sometimes restrict them.

In this sense, the agri-food routes emerge as a strategy of cultural valorization of great socioeconomic importance for the territories whose systems coexist around products with territorial anchorages through intersections between the sectors of gastronomy, agriculture, tourist services, and cultural heritage. The sustainability paradigm is positioned as a crucial issue in local public agendas. However, it sometimes constitutes points of conflict between the actors, who, as economic subjects, tend to conduct themselves under a productive business logic, seeking profit maximization to the detriment of other elements of sustainability, such as social development and environmental welfare.

As could be seen in this paper, the route is positioning itself as a project that offers food goods and tourist services, demonstrating great potential for territorial development, but at the same time awakening internal (communities and residents) and external (tourism and foreign business people) interests. In addition, other power groups, whose interests generate disagreements on the inequitable distribution of the benefits related to the use of territorial resources are affecting the territory.

This is due to opportunistic conduct, such as the creation of a Regulatory Council of Yahualica Chili, which takes advantage of legal loopholes, but is outside the consensus of international protocols and guidelines frameworks (such as the Mexican Standard), generating confusion among all the institutions that regulate and manage the DO.

This has slowed down the institutionalization of the route, causing more uncertainties about its viability to achieve rural development of the territory where the DO extends, through a balance between the ecological, cultural, economic, and productive subsystems.

References

¿Dónde crecen los chiles en México? In A. Aguilar, A. Lira, A. Aguilar, et al. (Eds.), Los chiles que le dan sabor al mundo. Contribuciones disciplinarias. México: Universidad Veracruzana, IRD, En.

Aguilar, A., et al. (2018). *Los chiles que le dan sabor al mundo. Contribuciones disciplinarias.* México: Universidad Veracruzana, IRD.

Appadurai, A. (2001). *La modernidad desbordada. Dimensiones culturales de la globalización.* Montevideo: Ediciones Trilce.

BBC News Redacción (2020). Glifosato: 3 preguntas sobre el herbicida por el que Bayer tendrá que pagar casi US$11.000 millones en demandas. *BBC Mundo Noticias*, 25 de junio. Disponible en: https://www.bbc.com/mundo/noticias-53180741. Consultado 20-11-20.

Brunori, G., & Rossi, A. (2000). Synergy and coherence through collective action: some insights from wine routes in Tuscany. *Sociologia Ruralis*, 40(10), 406–423.

Carton de Grammont, H. (2004). La nueva ruralidad en América Latina. *Revista Mexicana de Sociología, 66*, 279–300.

Centro Regional de Servicios Integrales para la Agricultura Protegida, CRESIAP. (2020). Reporte de Estudios Estratégicos y de Factibilidad para el Instituto del Chile Jalisco *(Documento inédito)*. México: CRESIAP, Índice Consultores, Instituto del Chile.

De Jesús-Contreras, D., & Medina, F. X. (2022). Reflexiones etnográficas sobre la descontextualización del patrimonio alimentario en el marco de las rutas gastronómicas en Querétaro (México). *Disparidades. Revista de Antropología*, 77(1). Available from https://doi.org/10.3989/dra.2022.013, accessed 22.09.22.

Delgado, A., & Hernández, E. (2019). Patrimonios inmateriales, desarrollo rural y despoblación. La identidad como recurso. *PH Boletín del Instituto Andaluz del Patrimonio Histórico*, 27(98), 150–171.

Espeitx, E. (2004). Patrimonio alimentario y turismo: una relación singular. *PASOS. Revista de Turismo y Patrimonio Cultural*, 2(2), 193–213.

García, B. (2017). El chile Yahualica, la historia de una denominación de origen. *El Ciudadano*, diciembre. Disponible en: https://elciudadanojalisco.mx/patrimonio/el-chile-yahualica-la-historia-de-una-denominacion-de-origen. Consultado 20.11.20.

Goodman, D. (2003). The quality turn and alternative food practices: reflections and agenda. *Journal of Rural Studies*, 19(1), 1–7.

Herrera, F. (2013). Enfoques y políticas de desarrollo rural en México: una revisión de su construcción institucional. *Gestión y Política Pública*, 22(1), 131–159.

Lash, S., & Urry, J. (1994). *Economies of signs and space.* Londres: SAGE Publications.

Medina, F. X. (2017). Reflexiones sobre el patrimonio y la alimentación desde las perspectivas cultural y turística. *Anales de Antropología*, 51(2), 106–113.

Melero, N. (2012). El paradigma crítico y los aportes de la investigación acción participativa en la transformación de la realidad social: un análisis desde las ciencias sociales. *Cuestiones Pedagógicas, 21,* 339–351.

Millán, G., Morales, E., & Pérez, L. M. (2011). Análisis del turismo gastronómico en la Provincia de Córdoba. *Tourism & Management Studies, 8,* 78–87.

Mintz, S. (2003). *Sabor a comida, sabor a libertad. Incursiones en la comida, la cultura y el pasado.* México: Conaculta-Ciesas.

Miranda, P. (2020). *Un chile único en el mundo se cultiva en Jalisco.* CONACYT prensa. Disponible en: https://centrosconacyt.mx/objeto/un-chile-unico-en-el-mundo-se-cultiva-en-jalisco/. Consultado 11.12.20.

Monllor, N. (2013). El nuevo paradigma agrosocial, futuro del nuevo campesinado emergente. *Polis (Bologna, Italy), 12*(34), 203–223.

Murgado, E., Torres, F., Parras, M., & Vega, M. (2012). El aceite de oliva como elemento nuclear para el desarrollo del turismo. En: Flavián, C. y Fandos, C. Coords. Turismo gastronómico, estrategias de marketing y experiencias de éxito. Zaragoza, España: Prensas Universitarias de Zaragoza.

Pensado, M., & Martínez, S. (2015). La estrategia de canasta de bienes territoriales y su repercusión local en empleos e ingresos. El caso de Xicotepec de Juárez, Puebla. *Ciencia y Tecnología Agropecuaria, 16*(2), 217–237.

Rodríguez, E. (2018). La diversidad genética de Capsicum annuum de México. In A. Aguilar, et al. (Eds.), Los chiles que le dan sabor al mundo. Contribuciones disciplinarias. México: Universidad Veracruzana, IRD, En.

Schejtman, A., & Berdegué, J. A. (2004). *Desarrollo territorial rural.* Santiago de Chile: RIMISP.

Secretaría de Agricultura y Desarrollo Rural (SADER) (2020). Servicio de Información Agroalimentaria y Pesquera (SIACON). Disponible en https://www.gob.mx/siap/documentos/siacon-ng-161430. Consultado 17.11.20.

Secretaría de Economía (SE) (2018). Declaración General de Protección de la Denominación de Origen "Yahualica." Disponible en http://dof.gob.mx/nota_detalle.php?codigo = 5516488&fecha = 16/03/2018. Consultado 10.11.20.

Secretaria de Medio Ambiente y Recursos Naturales (SEMARNAT), REq Consultores, Gobierno del Estado de Jalisco (GJAL). (2018). *Programa de Ordenamiento Ecológico Regional de la Junta Intermunicipal del Medio Ambiente. Altos Sur del Estado de Jalisco (JIAS). Caracterización.* (Documento inédito). Disponible en https://semadet.jalisco.gob.mx/sites/semadet.jalisco.gob.mx/files/4_agenda_ambiental_y_carecacterizacion_1_altos_sur_1.1.pdf. Consultado 11.12.20.

Suremain de, C. E. (2017). Cuando la alimentación se hace patrimonio. Rutas gastronómicas, globalización y desarrollo local. *TRACE, 72,* 165–181.

Thomé, H., Renard, M. C., Nava, G., & de Souza, A. (2014). La ruta del nopal (Opuntia spp.). Turismo rural y reestructuración productiva en el suelo rural de la Ciudad de México. *Rosa Dos Ventos, 6*(3), 390–408.

Tresserras, J., & Medina, F. X. (Eds.), (2007). *Patrimonio gastronómico y turismo cultural en el Mediterráneo.* Barcelona: Universitat de Barcelona-IEMed.

Yin, R. (2003). *Case study research.* Design and methods. Londres: SAGE Editors.

Zambrano, F., Trujillo, E., & Sólorzano, C. (2015). Desarrollo rural sostenible: una necesidad para la seguridad agroalimentaria en Venezuela. *Revista de Investigación, Administración e Ingeniería, 3*(1), 27–33.

CHAPTER ELEVEN

Gastronomic tourism and alternative food networks: a contribution to the Agenda 2030

María del Pilar Leal-Londoño
CETT- Barcelona School of Tourism, Hospitality & Gastronomy, Universidad de Barcelona, Barcelona, Spain

Introduction

In 2015 the United Nations General Assembly approved the Agenda 2030 for sustainable development and the sustainable development goals (SDGs). The framework includes 17 goals and 169 targets for achieving the goals, through which the Member States, the civil society, and the private sector could guide and measure their contributions to sustainable development until 2030.

This agenda balances the contributions to sustainable development regardless of the stakeholders (Spanish Global Compact Network Red Española del Pacto Mundial, 2016). According to the document, "The tourism sector and the Sustainable Development Goals" (2016), prepared with the participation of 13 companies in the tourism sector that identified potential challenges in the sector in relation to the new Agenda 2030, tourism can contribute, directly or indirectly, to all the SDGs. Specifically tourism appears in Goals 8, 12, and 14, which are related to inclusive and sustainable economic development, sustainable consumption and production, and the sustainable use of the oceans and marine resources, respectively. Despite this, and due to the transversality of the tourism sector, the contribution to the fulfillment of the SDGs by the private sector is also transversal to all the SDGs.

In the same document, emphasis was placed on the need to face the challenges of meeting the SDGs, specifically for tourism. However, when it comes to gastronomic tourism, it is important to understand the huge challenge that supposes merging tourism and food, two complex sectors that comprise a broad variety of interests. In this regard, it is worth reminding that the current global food production is characterized by industrialization and geographical concentration of production, with a few large-scale actors having power over a considerable share of overall food production (Gössling & Hall, 2013:13). The powerful intermediation of global multinationals decontextualizes food and turns customers away from any sort of reference to their geographic or social roots (Parrot et al., 2002). This fact is linked with the successive crises in food security registered in recent decades (as mentioned by several authors: Ilbery & Kneafsey, 1998; Ilbery & Maye, 2005; Renting et al., 2003; Sánchez, 2009), which establishes at the same time, the foundations for the emergence of alternative food networks (AFNs), which are diametrically opposed to industrial ways of food supply and commercialization systems.

The main aim of this chapter is to discuss the gastronomic tourism supply chain by applying the concept of networks within the framework of the Agenda 2030. The network concept reflects the relational forms of thinking, which influences many agrifood studies, providing a dynamic character (Kneafsey et al., 2008). The chapter is structured as follows: the first part presents the conceptual aspects associated with gastronomic tourism, AFNs, and the link between gastronomic tourism and the Agenda 2030. In the following section, a case study is presented that discusses the reality of a geographically isolated area such as the Pyrenees in Catalonia. The subsequent section presents the main results based on the information gathered and analyzed, and, finally, the conclusions discuss the achievement of the chapter's aim.

Gastronomic tourism and alternative food networks

Tasting the gastronomic products of a place is a growing trend in modern tourism businesses. For many tourists, it is a search for pleasure and one of the main motivations for tourism (Kumar Dixit, 2019). In many cases, people dedicate days of their trip to experiencing gastronomy, in the same way that they look for other different elements of culture such as art, music, or architecture. Gastronomy plays a crucial role for destinations because it becomes an axis of attraction on which values, culture, identity, landscape, and people intersect. As such gastronomy has become a powerful tool in contributing to the sustainability of tourism.

According to Kumar Dixit (2019) gastronomic tourism research can be divided into three perspectives: the consumer, the producer, and the development of the destination. The characteristics of the three perspectives—the multitude of interests and the possible conflicts with which they have to deal—are considered (Anderson et al., 2017). From the perspective of the consumer, food, and cuisine can be a reason for travel, which is based on a search for an extraordinary experience and a combination of factors that influence the visit to destinations. On the other hand, the needs of food tourists are supplied and satisfied by the producers, which include food producers, distributors, hotels, restaurants, food markets, attractions, and food festivals (Hall & Sharples, 2003).

Undoubtedly, gastronomic tourism has grown considerably and takes on many different forms depending on the different places in which it manifests itself (Kumar Dixit, 2019), such as events and festivals focusing on food and beverages, food and beverage markets, typical and traditional dishes, food producers that develop tourism activities to promote their products, farms addressed to tourists, food-themed hotels and resorts, or food and drink souvenirs. However, this diversity brings complexity in terms of how food arrives to tourists if it manifests itself in many different forms. In the context of increased experiential tourism, face-to-face contact relations among local producers and consumers, or hosts and guests, are helping to create a framework for authentic experiences. This is often associated with the fact that interaction, participation, and involvement in various activities create more positive customer feelings in various settings and situations, and thus enhance the value for the participants (Prebensen & Foss, 2011). Kim et al. (2010) argue that involvement in host community activities, which in gastronomic tourism could be local food events or picking up your own fruits on local farms, leads to a memorable tourist experience.

We should assume that tourism in relation to host-guest produces a space that allows the exchange of food products. At the same time a number of values shared by the various actors can be identified, including values centered on a concern for the environment, such as organic production; economic values, such as small-scale production, and social and cultural values, such as preserving traditional production methods (Leal-Londoño, 2015). All of these might be considered part of the sustainable framework presented in the Agenda 2030. The relationship between tourism and gastronomy offers a platform for the promotion of cultures, through culinary resources (Kumar Dixit, 2019) while contributing to sustainable territorial development (Leal-Londoño, 2019). Gastronomy contributes positively to many levels of the tourism value chain, such as agriculture, culture, and local food. Therefore it is not surprising that gastronomy tourism is currently considered a form of sustainable tourism.

Sustainability in the context of food is linked to slow food, Km 0, zero waste, local, and proximity food production, among other practices that are focused on reducing the impact on the environment and at the same time reinforcing local cultures. In this context, the concept of AFNs emerges as a representative form of a postproductivist production regime (Ilbery & Bowler, 1998; Renting et al., 2003) away from agroindustrial production.

The AFN refers to face-to-face interactions, such as farmers' markets, proximity relationships that transport the local and regional identity of a commodity, and the extended relationships that shorten the trading space sometimes like fair trade networks (Whatmore, Stassart, & Renting, 2003). Authors like Gössling and Hall (2013) call it new culinary systems that emerged from food citizenship and social justice in the food supply chain. In general, AFNs allow consumers to (perceived) fresher, safer, and tastier foods whose origin is known and trusted (Paül et al., 2013).

In contrast to the conventional chain, in the alternative chains that make up the AFNs there is a direct interaction not only with the producers or suppliers but with the whole chain. In the chain of alternative values an economic and ecological dimension coexist; that is to say, aside from an economic value, ecological values are also taken into account and are shared throughout the chain with other actors on the basis of cooperation and communication (Morgan et al., 2006). These aspects are reflected in the permanent search in tourism for a closer relationship between producer and final consumer (Antonioli Corigliano, 2002; Boniface, 2003; Hall & Sharples, 2003; Hjalager & Richards, 2002).

According to Jarosz (2008), generally AFNs may identify in four major ways: (1) shorter distances between producers and consumers; (2) small farm size and scale or environmental farming; (3) existence of alternative food purchasing models and venues, usually based on human relationships and proximity between consumers, producers, and/or retailers; and (4) commitment to the social, economic, and environmental dimensions of sustainable food production, distribution, and consumption.

Managing sustainability means involving all the actors in the local system (operators, authorities, tourists, residents), who usually have different interests and pursue various goals that usually do not coincide (Corvo & Filipo, 2019). According to Corvo and Filipo (2019) sustainable gastronomic tourism is based on the acceptance of general principles of behavior by clients, institutions, operators in the sector, and the local population. It also leads to a series of paradoxes that negatively risk the environment and the resources of the communities that this form of tourism tries to preserve. In this regard gastronomic tourism represents an opportunity to approach tourism in a sustainable way. The following section discusses the link between gastronomic tourism within the framework of the Agenda 2030.

Gastronomic tourism and the Agenda 2030

The World Tourism Organization defined sustainable tourism in 2005 as tourism that "takes full account of current and future economic, social, and environmental impacts to meet the needs of visitors, industry, the environment, and host communities." The conception of sustainability and sustainable development has evolved over the years. The concept of sustainable gastronomic tourism encompasses the need to preserve traditions, protect agriculture that rescues local varieties, promote the agricultural landscape and its customs, and value what food represents in a broad sense for a certain territory.

The Agenda 2030 for sustainable development and the 17 SDGs emphasize the contributions to sustainable development regardless of the stakeholders that make them (Spanish Global Compact Network, 2016). Tourism can contribute, directly or indirectly, to all the SDGs but specifically, tourism appears in the Targets of Goals 8, 12, and 14, which are respectively related to inclusive and sustainable economic development, sustainable consumption and production, and the sustainable use of the oceans and marine resources."

If the priority is to locally produce and sustainably grow fruits, vegetables, and seafood then these strategies could make a significant contribution to the SDGs (Scheyvens & Laeis, 2021). Furthermore, by shortening the food supply chain it is also possible to reduce energy use for transportation and refrigeration, which saves food miles (Gossling et al., 2011), contributing to SDG 13 in the fight against climate change or SDG1 aimed at improving the prospects for local development by eliminating poverty, especially in developing countries.

From the perspective of the tourists, local food provides more than sustenance: it is seen as an important part of the tourist experience, presenting an opportunity to consume a supposedly authentic aspect of the host's culture (Hall & Sharples, 2003). According to Scheyvens and Hughes (2015), the interest in local food is mainly due to a growing concern among western consumers about the sustainability of their food choices. If tourists seek sustainable practices during their trips that can contribute to the goals set out in the Agenda 2030 and its SDGs. Gastronomic tourism does incorporate sustainable practices unrelated to agribusiness and that converge toward an alternative agricultural model.

The Pyrenees tourism brand: a case study

The objective of this chapter focuses on analyzing the contribution of supply-demand exchange networks that are generated within the framework of gastronomic tourism and their contribution to sustainable development within the framework of the Agenda 2030. To achieve the goal, the gastronomic actions developed by the tourism brand Pyrenees located in Catalonia is taken as a case study. The choice of the Pyrenees brand lies in its geographical location, which makes it a geographically marginal area that corresponds to what is located in Catalonia. This geographical isolation contributes to the study to the extent that it allows assessing the contribution of gastronomic tourism in territories whose tourist vocation is not focused on sea, sun, and beach tourism. Its contribution to sustainable development could have a greater impact on places with similar characteristics.

The Pyrenees tourism brand is made up of 10 counties that extend over the long northern strip of Catalonia formed by the Pyrenees and the Pre-Pyrenees. Its tourist offer is clearly oriented toward hiking with more than 1000 km of marked trails. This is complemented by snow tourism (skiing), mountain biking, fishing, hot springs, and golf. Activities related to adventure tourism are currently being developed (rafting, paragliding, skydiving, among others); without neglecting cultural tourism and fairs around gastronomy. In 2020 the Pyrenees tourism brand received a total of 159.000 travelers for rural tourism according to the data from the Statistical Institute of Catalonia (IDESCAT). This data positions the brand as the first rural tourism destination in Catalonia.

With the aim of assessing the contribution to the sustainable development of gastronomic tourism in the territories that make up the Pyrenees brand, Goal 12 was specifically taken into account, focusing on responsible production and consumption. According to the United Nations, sustainable consumption and production consist of doing more and better with less, as well as it is a matter of decoupling economic growth from environmental degradation, increasing resource efficiency, and promoting sustainable lifestyles. Sustainable consumption and production can also make a substantial contribution to poverty alleviation and the transition to green and low-carbon economies.

Considering Goal 12, the analysis applied to gastronomic tourism is based on the scientific literature compiled by Leal-Londoño (2013) on contemporary elements that contribute to sustainable development at the territorial level. These elements incorporate the environmental, social, cultural, and economic dimensions necessary to establish a balance on the path toward territorial sustainability based on responsible production and consumption within gastronomic tourism (see Table 11.1).

To analyze the case study, a telephonic survey of 31 farmers located in the Pyrenees region, 25 intermediaries, and 22 restaurants in the area chosen from a public database of the Generalitat de Catalunya called Gastroteca was conducted. The main objective of the survey was to learn about the processes of integration and cooperation in the value chain of gastronomic tourism, participation in gastronomic associations, and, in general, the values shared among the stakeholders in such a way that sustainable aspects could be identified in the social, economic, cultural, and environmental dimensions that could be interrelated with SDG 12. The information gathered from the survey was systematized and subsequently analyzed through descriptive statistical analysis in Excel.

After the survey the results obtained for the case study, considering the concepts around AFN where personal contact is relevant for building trust and allowing exchange of food products within the food supply chain, 58.2% of restaurants contact their suppliers personally, while the personal contact of the intermediaries represents 28.7%. In the case of farmers, personal contact prevails for 18.3%. However, for farmers and intermediaries, the second most common means of contact with their suppliers is telephone contact, with 48.3% of the farmers surveyed mentioning it. While for 31.7% of the intermediaries it is their main means of contact. This fact confirms the importance of personal relationships and the trust established between stakeholders that has an impact on the relationship between food

Table 11.1 Key elements linked to territorial sustainability due to gastronomic tourism. Source: Leal-Londoño (2013).

Key elements linked to sustainability	Gastronomic attributes	Criteria
Integral: Integrates all sectors	Producers, marketers, restaurateurs/hotels	Takes into account all stakeholders and sectors involved in tourism activity
Endogenous: Must have territorial, cultural, and historical resources that are associated with agrifood products	Social-based gastronomic associations that promote local products	They are popular-based constructions interested in gastronomy
Balanced: There are inclusive proposals for the minority population.	Artisan businesses that operate based on local and proximity products	For being a minority in the agrifood industry
Popular based: Gastronomic strategies are open to anyone who wants to participate	Gastronomic fairs and events that promote identity based on local products	There is a high level of participation of all kinds of stakeholders, both public and private.
Cooperative: There is the presence of these culinary cooperation networks that rescue local varieties and culinary traditions (cuisine collectives)[a]	Cooperative and socially based cuisine collectives that promote local and traditional products and rescues local varieties	There is a cooperative spirit in these collectives.
Ecologic: The environmental value predominates as an element of exchange between tourists and farmers	Quality brands such AO, GPI, and Q Brand of Catalonia	These brands include, in many cases, greez or environmentally sustainable values.
Sociocultural: Agrofood products are considered heritage	Museums and rural heritage Fairs and events	These resources reinforce the local culture and in general that of local food

[a]The cuisine collectives are an association of chefs who promote local cuisine and the culinary traditions of a certain territory. It is a peculiar manifestation of Catalonia.

tourists and the destination. This supports the view mentioned by other authors in the analysis of the AAF, such as those carried out by Marsden et al. (2000), Whatmore et al. (2003), or Watts et al. (2007) regarding the relevance of personal contact.

According to Grieve and Slee (2003), the reconnection between producers and consumers within the food supply chain serves to rebuild relationships of trust between consumers. Activities such as farmers' markets and other forms of direct marketing help to increase dialog between the two parties. Indeed, these new scenarios have been highlighted for their potential as tourist attractions (Boniface, 2003; Hall & Mitchell, 2001; Hjalager & Richards, 2002; Ilbery & Kneafsey, 1998; Schlüter, 2009).

On the other hand, when asked about the link of each of the groups surveyed in the organizations, such as cuisine collectives or gastronomic routes, the three groups mostly bet on belonging to cuisine collectives (47.2%) as well as the farmers (50%) and the restaurants, due to the visibility that this type of organization gives to them.

It is worth mentioning that, in this territory, the production, processing, and consumption of meat are relevant not only today but also historically. This is how various products are recognized for their differentiated organoleptic properties, in some cases resulting from traditional and artisanal production methods, achieving their certification through appellations of origin (AO), protected geographical indications (PGI), or quality brands, such as the Q Brand, among others. An example is beef from the Catalan Pyrenees.

It is worth highlighting that for all the farmers/processors surveyed, it was found that the majority (90%) have a store selling directly to the consumer, especially the farmers who are dedicated to the production or processing of meat and cheese. This demonstrates high degrees of transformation of the product that allows it to reach the final consumer, thus shortening the supply and distribution chain and responding to the characteristics of short chains immersed in AFNs.

The fact that their suppliers come from neighboring municipalities (35.3%) or neighboring counties (39.5%), denotes the possibility of supplying mainly in intercounty areas, an aspect that can contribute to strengthening such trends as the consumption of Km 0 products. Taking into account the attributes identified in Table 11.1 and after applying them to the case study of the Pyrenees brand based on the information gathered, it was observed that it is possible to find examples for each of the features associated with sustainability and the contributions that gastronomic tourism can make to the sustainable local development of the Pyrenees brand (see Table 11.2).

When analyzing the contribution of the actions carried out in terms of gastronomy by the Pyrenees brand to SDG 12, it is important to highlight that products such as organic beef from the Pyrenees, beers, and organic artisan cheeses are part of the offer of the tourist brand. Undoubtedly, these products are attractive to tourists interested in gastronomy who can buy them as souvenirs among other purchase motivations. Moreover, the existence of a food fair dedicated to organic products is identified, which denotes awareness and willingness to respect the environment that dates back to the year 1048. It is worth mentioning that food markets are considered gastronomic tourist attractions and more importantly are spaces for exchange not only of products but also for relationships in terms of knowing the otherness. For Smith and Xiao (2008), farmers' markets, restaurants, and food festivals are part of gastronomic tourism; however, for Hall and Page (2006) the direct sale, makes up the morphology of the supply of this tourist modality. In any case, farmers' markets, direct sales, and tourism products are certainly alternative spaces where consumers or tourists are directly connected to producers and allow the establishment of proximity relations, one of the main features of AFNs.

Discussion and conclusions

The main purpose of this chapter is to explore the contribution of gastronomic tourism to sustainable tourism by analyzing the exchange relationships between agrifood products that are produced within the framework of gastronomic tourism. The chapter proposes a case study that represents a geographically isolated area. The Pyrenees as a case study is of special interest for tourism and for analysis since it can serve as an example of territories with similar spatial elements. Based on the concept of sustainable tourism and the 2030 Agenda as a global framework where tourism finds clear guidelines to the extent that goals are set that materialize the possible contributions of tourism, an analysis of the product exchange relationships that illustrate production and consumption was done. The main aim was to know the contributions to specific SDGs to investigate the contribution of gastronomic tourism to sustainable tourism within the framework of the Agenda 2030.

Table 11.2 Key elements linked to territorial sustainability due to gastronomic tourism in the Pyrenees tourism brand.

Key elements linked to sustainability	Gastronomic attributes	Criteria
Integral: Integrates all sectors	Producers, intermediaries restaurants/hotels	348 farmers 95 intermediaries 89 Restaurants 1 Michelin star restaurant
Endogenous: It must have territorial, cultural, and historical resources that are associated with agrifood products	Social-based gastronomic associations that promote local products	1 gastronomic association
Balanced: There are inclusive proposals for the minority population.	Artisan businesses that operate based on local and proximity products	AO Beef from the Pyrenees The Cheese of Alt Urgell and Cerdanya Butter from Alt Urgell and Cerdanya The Fesols of Santa Pau Bruneta from the Catalan Pyrenees
Popular based: Gastronomic strategies are open to anyone who wants to participate	Gastronomic fairs and events that promote identity based on local products	Welcome to Pagès Chestnut and Panellets Fair of the Girella del Pont de Suert
Cooperative: There is the presence of these culinary cooperation networks that rescue local varieties and culinary traditions (cuisine collectives)[a]	Cooperative and socially based cuisine collectives that promote local products, traditions and rescue local varieties	8 Cuisine collectives

(Continued)

Table 11.2 (Continued)

Key elements linked to sustainability	Gastronomic attributes	Criteria
Ecologic: The environmental value predominates as an element of Exchange between tourists and farmers	Quality brands such AO, GPI, Q Brand of Catalonia	AO Beef from the Pyrenees Bruneta from the Catalan Pyrenees
Socio-cultural:Agro-food products are considered heritage	Museums and rural heritage Fairs and events	Tastet and Bolet Fair The Bagà Rice Festival Fair of Sant Ermengol

[a]The cuisine collectives are an association of chefs who promote local cuisine and the culinary traditions of a certain territory. It is a peculiar manifestation of Catalonia.

The chapter also explores gastronomic tourism within the basic features of AFNs and draws theoretical and practical insights. It might be argued that gastronomic tourism is creating alternative food supply chains, which go beyond the traditional value chain and are based on ecological or green values that allow for food products exchange. Tourists and final customers are able to gather product information directly from producers and farmers and taste food products without intermediaries. Gastronomic tourism is helping to create new realities in the exchange of food products and is helping to redistribute the economic value of the agrifood sector. In this sense, gastronomic tourism is more than a tourism modality. It is transforming the food supply chain relationships among farmers, producers, and customers within tourism.

Based on the results, it was observed that, in the interaction between farmers, intermediaries, and restaurants, trust is considered important because it allows for maintaining long-term relationships between suppliers. These relationships are strengthened due to geographical proximity and personal relationships usually established through face-to-face and telephonic contact. This type of personal relationship has an impact on territorial sustainability because it allows supply chains to be shorter and to talk about networks because it is understood that exchange is much more complex than a mere commercial exchange. In addition, trust stabilizes the relationship between the actors and implies a cooperative attitude between them. Trust allows farmers to buy their supplies in municipalities close to their place of production, thus contributing, among other aspects, to the reduction of the carbon footprint of the products.

PGI undoubtedly contributes to the preservation of native breeds of animals, preserving biodiversity and contributing to the ecological balance of the environment. The concept of AFNs allows us to understand the complexity of those who participate in the exchange of food products within the framework of gastronomic tourism.

The case study suggests that in contrast with the conventional food systems, the conventions that characterize the farmers, retailers, and restaurants are strongly linked with green or ecological values. Gastronomic tourism and its AFNs may be seen as a new reconnection among farmers and tourists that allow new interaction spaces and scenarios beyond the conventional food supply system.

Finally, the case study for gastronomic tourism can contribute to sustainable local development through the gastronomic attributes identified. However, an increase in the number of stakeholders participating in the development of gastronomic activities that value the gastronomic heritage of the territory is suggested. In the same way, it is identified that the Pyrenees tourist brand places little value on the gastronomic products it has. Therefore an opportunity for improvement is observed around the joint work between the tourist areas and the agriculture areas.

The chapter has analyzed one tourist brand for the case of Catalonia. However, as recommendations for future research, it is suggested to analyze other brands and other objectives in such a way that the territorial comparison can be extended in which the results can be validated or in any case compared with other geographical contexts.

References

Anderson, T. D., Musberg, L., & Therkelsen, A. (2017). Food and tourism synergies: Perspectives on consumption, production and destination development. *Scandinavian Journal of Hospitality and Tourism*, *17*(2), 1—8.

Antonioli Corigliano, R. (2002). The route to quality: Italian gastronomy networks in operation. In A. Hjalager, & G. Richards (Eds.), *Tourism and gastronomy* (pp. 166—185). London: Routledge.

Boniface, P. (2003). Tasting tourism: Travelling for food and drink. Burlington: Ashgate.

Corvo, P., & Filipo, M. (2019). Sustainable gastronomic tourism. In S. Kumar Dixit (Ed.), *The Routledge handbook of gastronomic tourism* (pp. 199—206). London: Routledge.

Gossling, S., Garrod, B., Aall, C., Hille, J., & Peeters, P. (2011). Food management in tourism: Reducing tourism's carbon footprint. *Tourism Management*, *32*(3), 534—543.

Gössling, S., & Hall, M. (2013). *Sustainable culinary systems: Local foods, innovation, and tourism & hospitality*. London: Routledge.

Grieve, J., & Slee, B. (2003). *Review of the local food sector in Scotland*. Edinburgh: NHS Heath Scotland.

Hall, C., & Page, S. (2006). *The geography of tourism and recreation: Environment, place and space*. London: Routledge.

Hall, C. M., & Mitchell, R. (2001). 'Wine and food tourism'. In N. Douglas, & R. Derrett (Eds.), *Special interest tourism: Context and cases* (pp. 307−329). Brisbane: John Wiley.

Hall, C. M., & Sharples, L. (2003). *Food tourism around the world: Development, management and markets*. Oxford: Butterworth Heinemann.

Hjalager, A., & Richards, G. (2002). Tourism and gastronomy. London: Routledge.

Ilbery, B., & Bowler, I. (1998). From agricultural productivism to post-productivism. In B. Ilbery (Ed.), *The geography of rural change* (pp. 57−84). London: Longman.

Ilbery, B., & Kneafsey, M. (1998). Product and place. *European Urban and Regional Studies [Online], SAGE Journals Database, 5*(4), 329−341. Available from http://www.journals.sagepub.com.

Ilbery, B., & Maye, D. (2005). Alternative (shorter) food supply chains and specialist livestock products in the Scottish English borders. *Environment and Planning A [Online], SAGE Journals Database, 37*(5), 823−844. Available from http://www.journals.sagepub.com.

Jarosz, L. (2008). The city in the country: Growing alternative food networks in Metropolitan areas. *Journal of Rural Studies, 24*(3), 231−244.

Kim, Y. H., Goh, B. K., & Yuan, J. (2010). Development of a multi-dimensional scale for measuring food tourist motivations. *Journal of Quality Assurance in Hospitality & Tourism, 11*(1), 56−71. Available from https://doi.org/10.1080/15280080903520568.

Kneafsey, M. R., Cox, R., Holloway, L., Dowler, E., Venn, L., & Tuomainen, H. (2008). *Reconnecting consumers, producers and food: exploring alternatives*. Oxford: Berg Publishers.

Kumar Dixit, S. (Ed.), (2019). *The Routledge handbook of gastronomic tourism*. London: Routledge.

Leal-Londoño, M.P. (2013). Turismo gastronómico y desarrollo local en Cataluña: El abastecimiento y comercialización de los productos alimenticios (tesis doctoral). Director Dr. Francisco López Palomeque. Universidad de Barcelona, Facultad de Geografía e Historia.

Leal-Londoño, M. P. (2015). *Turismo Gastronómico, impulsor del comercio de proximidad*. Barcelona: Editorial UOC.

Leal-Londoño, M. P. (2019). Alternative foodnetworks and gastronomy. In S. Kumar Dixit (Ed.), *The Routledge handbook of gastronomic tourism* (pp. 516−526). London: Routledge.

Marsden, T. K., Banks, J., & Bristow, G. (2000). Food supply chain approaches: Exploring their role in rural development. *Sociologia Ruralis, 40*, 424−438.

Morgan, K., Marsden, T., & Murdoch, J. (2006). *Worlds of food: Place, power, and provenance in the food chain*. Oxford: Oxford University Press.

Parrot, N., Wilson, N., & Murdoch, J. (2002). "Spatializing quality: Regional production and the alternative geography of food.". *European Urban and Regional, 9*, 241−261.

Paül, V., Mckenzei, F., Araujo, N., & Rodill, X. (2013). Alternative food networks or agritourism? The 'Vegetable Tourism' experience in the Barcelona Peri-Urban area. In A. Firmino, & Y. Ichikawa (Eds.), *CSRS 2013*. Proceeding of the 21st colloquium of the commission on the sustainability of rural systems (pp. 114−128). Nagoya: International Geographical Union (IGU).

Prebensen, N. K., & Foss, L. (2011). Coping and co-creating in tourist experiences. *International Journal of Tourism Research, 13*(1), 54−67.

Renting, H., Marsden, T., & Banks, J. (2003). Understanding alternative food networks: Exploring the role of short food supply chains in rural development. *Environment and Planning A, 35,* 393−411.

Sánchez, J. L. (2009). Redes alimentarias alternativas: Concepto, tipología y adecuación a la realidad española. *Boletín de la AGE, 49,* 185−207.

Scheyvens, R., & Hughes, E. (2015). Tourism and CSR in the Pacific. In S. Pratt, & D. Harrison (Eds.), *Tourism in Pacific Islands* (pp. 134−147). New York, NY: Routledge.

Scheyvens, R., & Laeis, G. (2021). Linkages between tourist resorts, local food production and the sustainable development goals. *Tourism Geographies, 23*(4), 787−809.

Schlüter, R. (2009). Turismo gastronómico y medio ambiente, en busca de la sostenibilidad. *Revista de Economía, sociedad, turismo y Medio Ambiente, 8−9,* 43−63.

Smith, S. L. J., & Xiao, H. (2008). 'Culinary tourism supply chains: A preliminary examination'. *Journal of Travel Research, 46*(3), 289−299.

Spanish Global Compact Network (Red Española del Pacto Mundial). (2016). In *El Sector Turístico y los Objetivos de Desarrollo Sostenible, Turismo Responsable un Compromiso de Todos.* (2016). Madrid: Naciones Unidas.

Watts, D., Ilbery, B., & Jones, G. (2007). "Networking practices among 'alternative' food producers in England's West Midland Region." In D. Maye, L. Holloway, & Kneafsey (Eds.), Alternative food geographies. Representation and *practice* (pp. 287−307). Amsterdam: Elsevier.

Whatmore, S., Stassart, P., & Renting, H. (2003). What's alternative about alternative food networks? *Environment and Planning A, 35*(3), 389−391.

CHAPTER TWELVE

Gastronomic tradition, sustainability, and development: an ethnographic perspective of gastronomy in Las Hurdes (Extremadura, Spain)[1]

David Conde-Caballero[1,2], Borja Rivero Jiménez[3] and Lorenzo Mariano-Juárez[1,2]

[1]Department of Nursing, Faculty of Nursing and Occupational Therapy, University of Extremadura, Caceres, Spain
[2]International Commission on the Anthropology of Food (ICAF), Caceres, Spain
[3]Department of Business Management and Sociology, University of Extremadura, Cáceres, Spain

Introduction: the construction of the "black legend" of Las Hurdes

The region of Las Hurdes is located in the northernmost area of the province of Cáceres—a province that, together with Badajoz, belongs to the Autonomous Community of Extremadura, on the Spanish western border with Portugal. The region is divided into seven municipalities with more than 40 rural hamlets or *alquerías*.[2] Las Hurdes is a land of high mountains, traversed by five meandering rivers, contained to the north by the Sierra de Francia and Las Batuecas Natural Park. For many centuries, its relative isolation—due to its rugged topography, the lack of communication routes, and the historical neglect of authorities—contributed to creating a legend of remoteness and inaccessibility, and the perception of Las Hurdes as an almost exotic land, forgotten by time and history.

---

[1] This research was funded by "Programa Operativo FEDER Extremadura (2014—2020) y Fondo Europeo Desarrollo Regional (FEDER)", grant number: GR21153.
[2] The region of Las Hurdes covers an area of approximately 500 km² and has a low population density—less than 7000 inhabitants, spread among seven main villages and 40 rural hamlets locally referred to by the Arabic name of *alquerías*.

This perception was crystallized by the Spanish film-maker Luis Buñuel[3] in his 1933 documentary *Land without Bread*.[4] This documentary, characterized by its political intentionality and heavy surrealist aesthetic, described a territory built upon deprivation and want, whose inhabitants, the impoverished peasants, lived in a state of permanent hunger—a land without bread in its broadest sense (Matías, 2017, 2020). Buñuel, who would go on to become a cult film-maker, used the extreme poverty of some of the dwellers of the local *alquerías*—captured in dramatic, visually shocking scenes—to create awareness and stir public consciousness about their plight. The images were accompanied by a running commentary delivered by a detached-sounding narrator, who dryly noted at one point: "Three girls eat a piece of stale bread soaked in water. Bread that, until very recently, was not known in Las Hurdes." Inspired by Maurice Legendre's thesis *Las Jurdes: Etude de Géographie Humaine* (1927),[5] the scenes in this "cinematic essay on human geography"—as Buñuel himself described it—shocked and disturbed viewers all over the world.

Buñuel's images and Legendre's thesis, however, were just the latest episode to be added to the "black legend" of this region's poverty—a legend that travelers, adventurers, writers, and researchers had extensively described since the end of the nineteenth century and beginning of the twentieth, even though many of these writers had never visited the region. The legend referred to an inhospitable, barren land unsuited for cultivation, inhabited by hungry peasants struggling for survival, who were perceived almost as subhuman beings—people who only ate "[. . .] chestnuts, potatoes, beans, and some fruit when in season and available. Who hardly ate meat, except when one of them [their goats] fell over a cliff and died" (Matías, 2017: 224). A land that, as the famous Spanish

---

[3] Luis Buñuel (1900–83) was a famous Spanish film-maker, considered one of the greatest and most influential of all time. Associated in his youth with the surrealist movement, after the Spanish Civil War he went into exile in France and Mexico to avoid Franco's censorship. His most famous films include *Un Chien Andalou* (1929), *The Forgotten Ones* (1950), *Viridiana* (1961), and *The Discreet Charm of the Bourgeoisie* (1972).

[4] Considered one of the best documentary films since, in 1964, the prestigious Mannheim Film Festival selected it as one of the 12 best documentaries ever filmed. It was screened for the first time in New York in 1941 and quickly acquired international fame. Although the documentary had been shot years before, in 1933, it was censored by the Spanish authorities at that time and withheld from public view.

[5] Maurice Legendre defended his doctoral thesis *Las Jurdes: Etude de Géographie Humaine* in Bordeaux (France) in 1927. This extensive ethnographic study soon became notorious—not only because of its content, but for revealing the existence of such a deprived, poverty-stricken region in the heart of Europe.

physician Gregorio Marañón stressed,[6] only presented "misery, anemia, goiter, cretinism [congenital hypothyroidism], and a horrible, Dantesque spectacle." Marañón even diagnosed an endemic illness in the area—"the Hurdes malaise": "The sick person starts to feel unwell mid-morning, after walking for a while the paths that lead him to his vegetable patch. The brief midday luncheon soothes the symptoms, which restart a couple of hours after this is finished" (Marañón [1922] 1994). Las Hurdes was thus perceived as a land of almost "wild people."[7]

Indeed, Gregorio Marañón's dramatic report of his journey to Las Hurdes forced the Spanish king, Alfonso XIII, to travel to the region to assess the situation first-hand. His historic visit, which took place in 1922 (this year marking its centenary anniversary), aimed to establish what was true and what was myth in everything that was said about the region. In his journey through Las Hurdes the king took in the harsh, rugged landscape covered in heather and rock rose flowers, but also the extreme poverty of its dwellers. A well-known anecdote describes how a local peasant offered him a stale, blackened loaf of bread. The king accepted the offering with a certain degree of surprise, enquiring: "And this, what do you eat it with?" To this the peasant replied with what had been, for many years, the main component of the Hurdano gastronomy: "With hunger" (Schmigalle, 1993). These testimonies contributed to creating a convoluted, prejudiced narrative in which fantasy and reality mixed—a black legend that turned Las Hurdes into a paradigm of misery, disease, and poverty. Since then, Las Hurdes has endured the consequences of this legend—with Hurdano people being used, as the Spanish anthropologist and historian Julio Caro Baroja pointed out, as an "exhibition of tremendism" (Barroso, 1985).

However, present-day Las Hurdes bears no resemblance to this near-mythical representation—a black legend built upon notions of hunger

---

[6] Gregorio Marañón was a Spanish physician, scientist, writer, and philosopher renowned for his scientific and historic works. Interested in the black legend of Las Hurdes, he traveled to the region in 1922 with a medical commission—a journey that changed the history of the region. Recognizing many examples of dramatic endocrine disorder, his studies explored the prevalence of endemic goiter and hypothyroidism among the local population, which he attributed to a lack of iodine in their diets caused by the geographic isolation of the area.

[7] It is worth noting, however, that there were also dissenting opinions. The Spanish writer and philosopher Miguel de Unamuno, who also journeyed to Las Hurdes, denied its black legend in his writings: "[The idea] that they bark, or almost, is a tall tale. They speak Spanish, and they speak it very well" (Unamuno 1922:117). "And always the complaints," as Unamuno also noted in his work describing his journey around the region. "The King should come here to eat what we eat."

and poverty. The graphic novel *Buñuel en el laberinto de las tortugas* [Buñuel in the labyrinth of the turtles][8] (Solís, 2008) offers a more modern and colorful portrait of that "land without bread" filmed by Buñuel. The animated film based on this novel, which won a Goya award from the Spanish Academy of Cinematic Arts and Sciences, or the documentary *Las Hurdes, tierra con alma* [Las Hurdes, land with a soul][9] have also attempted to overcome that biased, exaggerated perception of the region. Even though Las Hurdes is still lacking in terms of infrastructures and services, the region has taken significant strides toward modernity, with a development focused on and stimulated by the touristic appeal of its natural and cultural heritage (Sánchez Martín & Rengifo Gallego, 2019).

A group of anthropologists from the Interdisciplinary Group for Studies on Society, Culture, and Health (GISCSA) at the University of Extremadura, Spain, have been examining for some time the impact of past episodes of hunger—real or mythical—on present-day dietary ideology, beliefs, and perceptions in the region (Rivero Jiménez, Caballero et al., 2020; Rivero-Jiménez et al., 2020). In particular, they have analyzed how these episodes have affected the development of the tourism industry, especially its gastronomy and hospitality sectors (Conde-Caballero et al., 2022; Rivero Jiménez, Ortiz, et al., 2020). Based on empirical materials obtained during their ethnographic study, this chapter examines the relationship between that land of misery and poverty whose image was exploited by so many in the past, and the development of Hurdano gastronomy in the present. To this end, this study focuses on the construction of key categories such as abundance, exquisiteness, or excellence, exploring how a sustainable gastronomy might be developed despite the burden of such a complex past. Considering that the deeply disturbing human experience of hunger can affect dietary habits in an area for generations, this research was conceived around one critical question: Is it possible, despite the pervasive impact of the legend of the poverty of Las Hurdes, to enhance the touristic appeal and economic growth of the region by implementing a sustainable gastronomic tourism model, without rejecting traditional cuisine?

---

[8] This graphic novel describes how Buñuel and his team experienced the filming of *Land without Bread*.

[9] This documentary from 2016 revisits Buñuel's film about Las Hurdes eighty years after it was originally filmed, taking a new look at the region.

Methods

This study is a qualitative, ethnographic analysis of dietary ideologies in the region of Las Hurdes (Extremadura, Spain) based on fieldwork conducted in the area since 2018. Ethnographic material regarding the development of tourism and gastronomy in the area was collected between February 2020 and March 2022. Our research approach is based on classic ethnographic fieldwork principles (Hammersley & Atkinson, 1994; Velasco & Díaz de Rada, 2006). The method has been previously applied in research literature exploring beliefs and ideologies in food and nutrition (Harris et al., 2009; Ottrey et al., 2018).

Data were gathered through in-depth, semistructured interviews (Kvale & Brinkmann, 2015; Neumann & Neumann, 2018). A total of ten interviews were conducted with restaurant managers and chefs in the region. The study also included data obtained through informal interviews and conversations with residents of the region, tourists, and government officials in different local institutions. The interview guide was designed around the main question underpinning the study, and it took into account our previous ethnographic research on food and nutritional issues (Rivero-Jiménez et al., 2020) and similar studies on this subject (Batat, 2021). At the same time, the research team was open to exploring any new themes not considered initially that could emerge during the interview process (Emans, 2019), thus adding nuance to the subject of study. The study also relied on informal conversations (Swain & Spire, 2020) with different local actors—i.e., tourists, restaurant guests, and restaurant staff—carried out in different observational units (Shah, 2017) such as eating/working areas within restaurants, gastronomic festivals, or local celebrations.

One of the main obstacles our study faced—particularly during the period between March and July 2020, but also in other moments until March 2022—was the disruption caused by the COVID-19 pandemic due to the restrictions that were placed on travel and the implementation of social distancing measures, all of which impacted our fieldwork schedule. In adapting to this situation, part of this work has been subjected to the challenges derived from implementing a virtual ethnography (Estalella & Ardévol, 2011; Estalella Fernández, 2018; Hine, 2011). The study also included virtual observational units (Achmad et al., 2020), through which we monitored comments and reviews in different Internet forums, blogs, and websites.

Abundance versus the narrative of deprivation

In any examination of present-day Las Hurdes, it soon becomes clear that the region is still struggling with the deep-rooted impact of its historic black legend. To counterbalance this negative image, the region is being promoted as a culinary touristic destination, harnessing its gastronomy to offset the prejudices about its deprivation and misery prevailing over the past centuries. Indeed, one of the main categories emerging from the empirical materials gathered during our fieldwork is the idea of abundance, suggesting perhaps a legacy of memories of hunger and want—and a need to compensate in the present for a past built upon deprivation. This idea of abundance is easily appreciable in the conspicuous consumption that underpins celebrations and festive occasions in present-day Las Hurdes. Food consumption—even excessive food consumption, on occasion—is a cornerstone of celebratory practices in a region where, in the past, food was only conspicuous because of its scarcity. The exuberance with which food is offered to visitors feels almost as if present-day Hurdanos are trying to detach themselves from the deprivation suffered in the past. Indeed, during our interviews and observations, it was rare not to be offered food and drink—often excessively so. As one of our participants described:

> "Many things have been lacking here. It is true that there has been no bread until recently. If you wanted bread you had to travel far, and it was impossible. But there was no lack of other products of the land. There have always been potatoes, and olives, and I don't know... everything. We were poor, but we were never hungry. We have always lived from the land — but that was before. Now we have anything you might want. Look, try this, try this. You'll see how [good] this is...."

However, food does not just underpin the locals' behavior toward visitors to the region—gastronomy is also a key component of how the area is being marketed as a touristic destination. There are other aspects of Las Hurdes that could have been used to seduce and attract visitors, such as the peculiar architecture so well described by Catani (1998)—indeed, its buildings are used as narrative nodes in Solís's storytelling (2008). Or perhaps the routes take travelers through an exceptional landscape whose lush watercourses never fail to surprise. However, a clear decision has been made to put food at the center of the picture—even prioritizing quantity before quality when showcasing gastronomy to visitors from outside the region. The frequent celebration of traditional *matanzas* during local festivities in different villages is a clear example of this intentionality. *Matanzas* are seasonal festivals during which a large number of pigs are slaughtered, their meat is used to prepare different traditional dishes or cured to be consumed at a later date. This kind of celebration pays homage to a crucial time of the year for Hurdano families—for those who could afford it, this seasonal pig slaughter would guarantee the provision of meat for the rest of the year. Nowadays, these ritual performances include tasting and sampling sessions with local dishes, wines, and liqueurs, which are generously offered to those attending.

Indeed, traditional *matanzas* are just one of many celebrations that put food consumption in the spotlight—presenting a clear contrast with a past shaped by experiences of deprivation and struggle for survival. For instance, visitors to the region can now follow different *tapas* routes and attend *tapas* festivals,[10] and there are also cultural and gastronomic routes focused on particular local products such as olive oil or honey—the International Beekeeping and Tourism Fair is focused on what is one of the main delicacies of the region, showcasing its production and derived products—or the Kid Goat Festival, the jewel of Extremadura's gastronomic crown.

All these examples explain why government institutional advertising campaigns—at the local, regional, and provincial levels—focus on food as a symbol of identity. At the same time, it helps Hurdanos reclaim their agency by providing an answer to a past image of the region that was created and exploited by others. For instance, Las Hurdes has been included in a catalog of Gastroexperiences backed by the government of the Autonomous Community of Extremadura—a compendium of touristic and gastronomic experiences intended to highlight the main attractions of the whole region. Likewise, traditional Hurdano recipes have been included in the government's Tentation/s program, which aims to advance and promote local gastronomy. Different villages also advertise local delicacies, while specialized blogs open their reviews of the region with firm statements such as "If you like your food, do not worry: in Las Hurdes you will not go hungry."

The food offered is always generous, abundant, and deeply rooted in tradition—the old-fashioned dishes that have always been prepared and

---

[10] *Tapas* are small plates of savory food that are served in different parts of Spain alongside drinks in bars and restaurants. In some places, particularly in southern Spain, they are usually offered for free when ordering a drink.

consumed in the region. It is interesting to note that Hurdanos have preserved the traditional ways of preparing local ingredients, instead of opting for what could have been an easier option—to distance themselves from traditional recipes because of their perceived association with poverty. However, instead of following global market trends, Hurdano restaurateurs have decided to reassert themselves through a commitment to local ingredients and traditions. Rather than feeling shame because of their past, as noted by Catani (1999), their response to the stigma has focused on their culinary heritage—what is theirs, what they do best. As a result, two overlapping categories are repeated in restaurants and gastronomic events: tradition and abundance in everything they offer.

Honey, pollen, cured meats, olives, cherries, chestnuts, and wild mushrooms—the same foods that were used in the past to define hungry, poverty-stricken Hurdanos—now sustain their sense of identity. This gastronomy has a strong personality, but the scars left by past times of deprivation are still noticeable in the importance given to quantity. At the same time, it is a cuisine that strives for excellence, and whose star dish is the grass-fed Hurdano kid goat, prepared in different ways: roasted, stewed, or cooked with pollen. Restaurant menus proudly offer local dishes that are both distinctive and enticing: "Hurdano lemon," a salad made with lemons, oranges, garlic, and chorizo sausage, topped with a fried egg, which used to be a traditional shepherd's breakfast; escarole salads; freshwater fish, served with *moje* or fried in local olive oil; typical Hurdano stews with evocative names such as *matajambres, cazuelas de rebujones, patatas meneás, pipos con berzas, habichuelas,* or *asaduras*; the *moje jurdano*, a strong and flavorsome garlic mayonnaise; sweet pastries such as *jugus curinus, piñonates, floretas, buñuelos,* or chestnut panna cotta; or artisan liqueurs such as honey *orujo*. These are just a few examples of typical dishes, always served in substantial portions.

Exquisiteness, sustainability, and new gastronomy

Although abundance in the servings and pride in traditional Hurdano recipes are the main identity traits of local gastronomy, a few new restaurateurs are also trying to explore new limits—bringing traditional ingredients and recipes in line with current times and the latest gastronomic trends. They tap into a niche defined by locally produced food and exquisiteness, aiming for refined, high-quality cuisine—in venues where even the aesthetics and interior design indicate a clean break with traditional restaurants. Although their chefs and managers proudly support the gastronomic characteristics and identity of the region, using its most distinct ingredients and flavors, they also focus on high-end presentation rather than abundance—which is a novel approach in this region. As the well-known Repsol tourist guide puts it, these new restaurants are a "breath of fresh air" with a "pinch of daring" thrown in, in so much as they not only focus on the category "tradition" but also on "exquisiteness" and "sustainability"—a model that aims to find a balance between economic growth and social well-being.

This new gastronomic approach, pushed by new generations of restaurateurs, has already succeeded in attracting a different kind of tourism to the region: people who look for authenticity, and do not mind traveling long distances or bad access roads to find it. One of the most successful of these novel restaurants, La Meancera, in the small *alquería* of El Gasco, is a good example of this new approach. Its renewed menu offers food with a very strong personality, negotiating the territory between tradition and "new palates": cured goat cheeses, cheeses in olive oil, kid goat cutlets with honey, *croquetas de prueba de cerdo* (made with pork marinated with garlic and paprika), or a renewed version of the traditional *patatas meneás (mashed potatoes seasoned with garlic and paprika)*. "Here, we look after traditional gastronomy, even though we innovate," states one of the chefs interviewed. As Fernández-Poyatos et al. (2019) suggest, this is a clear example of so-called "local produce cuisine," with its focus on the excellence of the ingredients used and its valorization of local, seasonal produce. Zero miles, farm-to-fork ingredients are a constant in this new narrative, as one of the interviewees stated:

> "From the producer to my kitchen, and from me to the diner guest. There are no intermediaries. We love our products — kid goat, pilonga chestnuts, acorns, free-range eggs, paprika, watercress, escaroles, pamprinas and emborujas[11], and of course, the aromatic herbs that are a gift from our mountains. And the honey."

These new venues are not just unashamed of their association with the region—rather, they are proud champions of traditional, locally sourced products. Their cuisine is based on flavors previously associated with hunger and misery, but they do not try to compensate for this with

---

[11] Small freshwater plants foraged for salads.

abundance. Rather, they use local ingredients to create zero miles, sustainable dishes—with local farms supplying the core of their menu. "Our menu changes with the season, using mostly zero miles ingredients and organic too, whenever possible; we also have our own kitchen garden, and during the summer our kitchen uses almost only local produce," noted the chef in one of these restaurants. This turn toward a sustainable diet (Buttriss & Riley, 2013) has an added value—it allows the survival of small businesses and farms that would be doomed to disappear in a more globalized supply chain model (Mauleón, 2010). The Hurdano model, instead, is firmly committed to turning gastronomy from abundance to sustainable exquisiteness.

In a trend also noticed in other regions (Santana, 2003), the Hurdano cultural heritage has been reinterpreted and reinvented to reach new potential clients. This new modernity looks for trusted, high-quality ingredients, but also includes values associated with a specific territory—its natural resources, its cultural heritage, and historical models of food production and preparation (Gutiérrez-García et al., 2020). The increasing number of restaurants following this international trend suggests that, once consolidated, culinary tourism could be used as a force for change in the promotion of poorer rural areas (Ab Karim & Chi, 2010). It could help improve the image of these regions as touristic destinations and stimulate their economic development (Albuquerque et al., 2019), eventually increasing the number of visitors to these areas (Jiménez Beltrán et al., 2016; Kivela & Crotts, 2006). Indeed, the potential of this new tourism model is already noticeable in Las Hurdes, as suggested by the increase in the number of visitors, as well as by a shift in their socio-economic profile. These are tourists ready for higher expenditure in exchange for a distinctive gastronomic experience based on high-end cuisine and ingredients with unique organoleptic properties (Serra et al., 2015). However, it is important to note that the combination of traditional and innovative features has to strike a certain balance (Dacin & Dacin, 2008), which has not yet happened in Las Hurdes.

Our research in the region revealed a new relationship between a past of deprivation and the tourism potential of the future. For many decades, the struggle for food in Las Hurdes resulted in a culinary tradition whose main aim was to demonstrate that they had enough food. By contrast, present-day Hurdanos are not worried about food insecurity, turning instead toward refinement and exquisiteness—a shift in line with the prevailing experiences sought in current tourism trends. The result of our

Gastronomic tradition, sustainability, and development

fieldwork in Las Hurdes suggests that, without neglecting the public attracted to the region for other reasons, authorities should also appreciate the opportunities offered by the promotion of an activity that could bring in a different kind of visitor. However, this would require enabling supportive policies (De Jesús-Contreras, Medina, 2021) and overcoming the traditional barriers to public—private partnerships (Medina, 2017). Previous research has suggested that these barriers could be overcome by implementing public policies that promote a gastronomic model based on quality and excellence (Castillo-Manzano et al., 2020), but there are also other possibilities—i.e., establishing cookery schools based on local gastronomy (Santich, 2004), providing specific training for workers in the hospitality sector (Romero et al., 2019), or adhering to initiatives stressing the uniqueness or authenticity of local culinary experiences (Patrimonio Gastronómico Protegido, 2020).

Final considerations

For many decades, the region of Las Hurdes has carried the stigma of its perception as a paradigm of extreme poverty and hunger. This study builds on an idea that the authors have explored in previous research—that the experience of food insecurity in the past can leave a lasting impact on communities, creating clear continuities in present-day nutritional choices and practices (Conde-Caballero et al., 2021). As noted in this study, this might also affect the gastronomic profile presented to tourists in these regions. The analysis of restaurants is particularly interesting, since the stigma associated with food insecurity could have negatively affected the perception of traditional food. Indeed, restaurateurs could have followed new culinary trends to signal a clean break from the past. However, in Las Hurdes their response in most cases has been the opposite—by committing to a cultural identity closely associated with their land and the consumption of traditional foods, they have faced misconceptions and stigma by championing the Hurdano culinary identity.

While overcoming a past marked by deprivation often meant overcompensating with a cuisine of abundance, in recent years a different approach has emerged—with new venues markedly committed to refinement and excellence. This new trend, with its focus on a sense of identity based on local ingredients, tradition, and sustainability, could also have a significant impact on touristic development. Indeed, this new stage in the evolution of Hurdano gastronomy could be a significant asset for the future economic and touristic development of the region—a region traditionally lacking in opportunities. As noted in other contexts, local and regional authorities could capitalize on this novel approach to promote tourism, thus contributing to the economic and social growth of an area where this is still much needed. At the same time, this approach could also help the region of Las Hurdes finally overcome its image as a land of hunger and deprivation—the land that time forgot.

References

Ab Karim, S., & Chi, C. G.-Q. (2010). Culinary tourism as a destination attraction: An empirical examination of destinations' food image. *Journal of Hospitality Marketing & Management, 19*(6), 531−555.

Achmad, Z. A., Ida, R., & Mustain, M. (2020). A virtual ethnography study: The role of cultural radios in campursari music proliferation in East Java. *ETNOSIA: Jurnal Etnografi Indonesia, 5*(2), 221−237.

Albuquerque, C. R. D., Mundet, L., & Aulet, S. (2019). The role of a high-quality restaurant in stimulating the creation and development of gastronomy tourism. *International Journal of Hospitality Management, 83,* 220−228.

Barroso, F. (1985). Los Moros y Sus Leyendas En La Serranía de Las Jurdes. *Revista de Folklore,* 50.

Batat, W. (2021). The role of luxury gastronomy in culinary tourism: An ethnographic study of michelin-starred restaurants in France. *International Journal of Tourism Research, 23*(2), 150−163.

Buttriss, J., & Riley, H. (2013). Sustainable diets: Harnessing the nutrition agenda. *Food Chemistry, 140*(3), 402−407. Available from https://doi.org/10.1016/j.foodchem. 2013.01.083.

Castillo-Manzano, J. I., Castro-Nuño, M., Lopez-Valpuesta, L., & Zarzoso, Á. (2020). Quality versus quantity: An assessment of the impact of michelin-starred restaurants on tourism in Spain. *Tourism Economics.* Available from https://doi.org/10.1177/ 1354816620917482.

Catani, M. (1998) Las Hurdes Desde Dentro y Desde Fuera. In: *Crónica Del II Congreso de Hurdanos y Hurdanófilos,* pp. 119−136.

Catani, M. (1999). Las Hurdes Como Imagen de Una Sociedad Local En Transformación. *Revista de Estudios Extremeños, 55*(2), 605−632.

Conde-Caballero, D., Rivero-Jimenez, B., & Mariano-Juarez, L. (2021). Memories of hunger, continuities, and food choices: An ethnography of the elderly in Extremadura (Spain). *Appetite, 164.* Available from https://doi.org/10.1016/j.appet.2021.105267.

Conde-Caballero, D., Rivero-Jiménez, B., & Mariano Juárez, L. (2022). De La Abundancia a La Exquisitez Gastronómica. Una Etnografía de Las Nuevas Oportunidades Para La Atracción Del Turismo En Las Hurdes (Cáceres). *PASOS. Revista de Turismo y Patrimonio Cultural, 20*(2), 453−464.

Dacin, M. T., & Dacin, P. A. (2008). Traditions as institutionalized practice: Implications for deinstitutionalization. *The Sage Handbook of Organizational Institutionalism, 327,* 352.

De Jesús-Contreras, D., & Medina, F. X. (2021). Turismo Gastronómico, Productos Agroalimentarios Típicos y Denominaciones de Origen. Posibilidades y Expectativas de Desarrollo En México. *Journal of Tourism and Heritage Research, 4*(1), 343−363.

Emans, B. (2019). *Interviewing: Theory, techniques and training*. Groningen/Houten: Routledge.

Estalella, A., & Ardévol, E. (2011). E-Research: Desafíos y Oportunidades Para Las Ciencias Sociales. *Convergencia, 18*(55), 87−111.

Estalella Fernández, A. (2018). Etnografía de Lo Digital: Remediaciones y Recursividad Del Método Antropológico. *AIBR: Revista de Antropología Iberoamericana, 13*(1), 45−68.

Fernández-Poyatos, M. D., Aguirregoitia, A., & Bringas, N. L. (2019). La Cocina de Producto: Seña de Identidad y Recurso de Comunicación En La Alta Restauración En España. *Revista Latina de Comunicación Social, 74*, 873−896.

Gutiérrez-García, L., Labrador-Moreno, J., Blanco-Salas, J., Javier Monago-Lozano, F., & Ruiz-Téllez, T. (2020). Food identities, biocultural knowledge and gender differences in the protected area 'Sierra Grande de Hornachos' (Extremadura, Spain). *International Journal of Environmental Research and Public Health, 17*(7). Available from https://doi.org/10.3390/ijerph17072283.

Hammersley, M., & Atkinson, P. (1994). *Etnografía. Métodos de Investigación*. Barcelona: Paidós.

Harris, J. E., Gleason, P. M., Sheean, P. M., Boushey, C., Beto, J. A., & Bruemmer, B. (2009). An introduction to qualitative research for food and nutrition professionals. *Journal of the American Dietetic Association, 109*(1), 80−90.

Hine, C. (2011). *Etnografía virtual*. Barcelona: Editorial UOC.

Jiménez Beltrán, J., López-Guzmán, T., & Santa-Cruz, F. G. (2016). Gastronomy and tourism: Profile and motivation of international tourism in the city of Córdoba, Spain. *Journal of Culinary Science & Technology, 14*(4), 347−362.

Kivela, J., & Crotts, J. C. (2006). Tourism and gastronomy: Gastronomy's influence on how tourists experience a destination. *Journal of Hospitality & Tourism Research, 30*(3), 354−377.

Kvale, S., & Brinkmann, S. (2015). *Interviews: Learning the craft of qualitative research interviewing*. London: SAGE Publications Inc.

Legendre, M. (1927). *Las Jurdes: Étude de Géographie Humaine*. Feret.

Marañón, G. (1994). *Viaje a Las Hurdes: El Manuscrito Inédito de Gregorio Marañon*. Madrid: El País-Aguilar.

Matías, D. (2017) La Producción Geosimbólica de Las Hurdes. Teoría, Historia y Práctica de Un Territorio Imaginario. Universidad de Extremadura.

Matías, D. (2020). La Leyenda de Las Hurdes. Geografía, Literatura e Historia de Una Comarca Mítica. Badajoz: Diputación de Badajoz.

Mauleón, J. (2010). Sistema Alimentari Sostenible i Agricultura Familiar: El Pasturatge Al País Basc. In F. X. Medina (Ed.), *Reflexions Sobre Les Alimentacions Contemporánies. De Las Biotecnologies Als Productes Ecológics* (pp. 73−110). Barcelona: Editorial UOC.

Medina, F. X. (2017). Reflexiones Sobre El Patrimonio y La Alimentación Desde Las Perspectivas Cultural y Turística. *Anales de Antropología, 51*, 106−113.

Neumann, C. B. B., & Neumann, I. B. (2018). *Interview techniques. Power, culture and situated research methodology* (pp. 63−77). Springer.

Ottrey, E., Jong, J., & Porter, J. (2018). Ethnography in nutrition and dietetics research: A systematic review. *Journal of the Academy of Nutrition and Dietetics, 118*(10), 1903−1942.

Patrimonio Gastronómico Protegido. (2020). Patrimonio Gastronómico Protegido.

Rivero Jiménez, B., Caballero, D. C., & Juárez, L. M. (2020). Educación Para La Salud y Alimentación En Personas Mayores. Tradición, Cultura y Autoridad En Las Hurdes (Extremadura-España). *Agathos, Atención Sociosanitaria y Bienestar, 3*, 18−25.

Rivero Jiménez, B., Ortiz, L. L.-L., Conde-Caballero, D., & Juárez, L. M. (2020). Del Hambre a La Excelencia. Aproximaciones Etnográficas a Las Hurdes Desde La Antropología de La Alimentación. *The Foodie Studies* (5).

Rivero-Jiménez, B., Conde-Caballero, D., & Mariano-Juárez, L. (2020). Health and nutritional beliefs and practices among rural elderly population: An ethnographic study in Western Spain. *International Journal of Environmental Research and Public Health*, *17*(16). Available from https://doi.org/10.3390/ijerph17165923.

Romero, I., Gómez, I. P., & Zabala-Iturriagagoitia, J. M. (2019). 'Cookpetition': Do restaurants coopete to innovate? *Tourism Economics*, *25*(6), 904–922.

Sánchez Martín, J. M., & Rengifo Gallego, J. I. (2019). Evolución Del Sector Turístico En La Extremadura Del Siglo XXI: Auge, Crisis y Recuperación. *Lurralde: Investigación y Espacio* (42), 19–50.

Santana, A. (2003). Mirando Culturas: La Antropología Del Turismo. In Á. R. Gil (Ed.), *Sociología Del Turismo* (pp. 103–126). Madrid: Ariel.

Santich, B. (2004). The study of gastronomy and its relevance to hospitality education and training. *International Journal of Hospitality Management*, *23*(1), 15–24.

Schmigalle, G. (1993). *Viaje a Las Hurdes: El Manuscrito Inédito de Gregorio Marañón y Las Fotografías de La Visita de Alfonso XIII*. Madird: El País Aguilar.

Serra, J., Correia, A., & Rodrigues, P. M. M. (2015). Tourist spending dynamics in the algarve: A cross-sectional analysis. *Tourism Economics*, *21*(3), 475–500.

Shah, A. (2017). Ethnography? Participant observation, a potentially revolutionary praxis. *HAU: Journal of Ethnographic Theory*, *7*(1), 45–59.

Solís, F. (2008). *Buñuel En El Laberinto de Las Tortugas*. Mérida: Editora Regional de Extremadura.

Swain, J. M., & Spire, Z. D. (2020). The role of informal conversations in generating data, and the ethical and methodological issues. *Forum Qualitative Sozialforschung/Forum: Qualitative Social Research*, *21*.

Unamuno, M. D. (1922). *Andanzas y Visiones Españolas*. Madrid: Renacimiento.

Velasco, H., & Díaz de Rada, Á. (2006). *La Lógica de La Investigación Etnográfica*. Madrid: Trotta.

CHAPTER THIRTEEN

Eating and thinking at the same time: food consumption and sustainability in Lugo (Galicia, Spain)

Elena Freire-Paz
Universidade de Santiago de Compostela, Lugo, España

> One day, I think, the intellectual monoculture of science will be replaced by a polyculture of complementary knowledge. And then there will be enough for everyone- (Robin Wall Kimmerer).

Introduction

As proposed by Claude Fischler in his book The (H)omnivore (Fischler, 1995), the omnivore nature of human beings involves a paradox: Although it allows them to reach a wider and more varied number of food resources, it results in a loss of specialization that makes it difficult for them to intuitively recognize the foods that are good for consumption from those that are not. The experience that constitutes social and cultural knowledge is in charge of fulfilling this function of discernment between valid and invalid foods. At the base of human nutrition, extended as a cultural product, is the action of thinking about food. The conceptual transposition of food from being considered a simple product or object to becoming food, that is to say, the mental transformation of a natural product into a cultural product. This is the ambiguity exposed by Claude Lévi-Strauss (1972) in terms of "good or bad for thinking" and "good or bad for eating," in the sense that cultures shape as valid and own certain alimentary uses, starting

---

☆ This research is part of the project *Novos modelos de hortas en Lugo e a súa contorna: saberes agroecolóxicos, seguridade alimentaria e resiliencia local* —NOMOHOLU— within the Collaboration Agreement between the Xunta de Galicia and the University of Santiago de Compostela, which regulates the Campus of Specialisation "Campus Terra".

from the consideration of the food as good to be thought of and, consequently, good to be eaten, simply from cultural criteria and regardless of its nutritional qualities.[1] Certainly, if we understand this appreciation to be valid, we should conclude that the degree of cultural appropriation that parts of society make of the food that they have classified as edible will be inversely proportional to their tolerance or acceptance of foreign uses and food products. Thus, the inclusion of a new type of food or a new culinary-alimentary use in a social group would previously need the consideration of this food to be thought of, that is to say, its inclusion in the mental framework of the social group. It is true that if we make a leap from the structuralist paradigm to cultural materialism, the causal orientation that is being detailed should be inverted, since, as Harris (1990) proposes, reversing the Lévi-Strauss formula, what is good to be eaten is also good to be thought of and sold. This determines that the selection of food by a community is strictly verified by material terms, for example, nutrition (whether it is nutritional or not), ecology[2] (whether it implies aggression toward the environment or not), economic facts (whether it implies economic facts or not), and historical facts. The best possible adaptation to the environment and not cultural arbitrariness is that which governs, according to this approach, the rules that explain the variability of food consumption.

In any case, the relation between the food that society uses and its ideological structures is indisputable because "the need to feed oneself, the procurement, purification, storage, processing, and consumption of food [...] necessarily expresses features of culture which, in terms of the invention of the everyday, harbor a fundamental cultural wealth" (Aguilar, 2014: 1). The justification of the current existence of foods labeled as organic or bio in the markets and the consumption habits of many western countries should be

---

[1] The paper Food, tradition and prestige,' De Garine (1976), analyzes prestigious foods among the *moussey, massa*, and *tupuri* tribes in Central Africa. In this regard, he comments that only human beings avoid the use of nutritionally valuable food because they are of 'low status' but they consume organoleptically mediocre and nutritionally poor products to appear economically prosperous (Contreras & Gracia, 2005: 266).

[2] An unavoidable reference when discussing ecological anthropology is Roy Rappaport, who at the beginning of the sixties, from a functionalist perspective, did field work in mount Bismarck in New Guinea, in a community of itinerant horticulturists, the *tsembaga*. In 1968 he published his book *Pigs for the ancestors. Ritual in the Ecology of a New Guinea People*, based on this experience. In this work, Rappaport explains how, through a complex cycle of rituals associated to the spirits of ancestors, the *tsembaga* maintain a balanced and ecological relationship with the physical environment where they develop their activities: through the control of the pig population, regulating the quantity of cultivated area, maintaining the fallow systems, the consumption of energy needed to subsist through the intake of proteins, etc.

based on some considerations of this type. Why is there an increased preference for these types of products especially in certain social sectors? What are the reasons for this choice? In general, beyond the theoretical dialectic established in anthropological studies[3] regarding the order in which the factors should be placed, the current consumption of organically produced food is linked to thought, insofar as it responds to a process of awareness, reflection, and assumption of a position not free of ethical values and moral considerations influenced by the more than worrying situation of the environment. This situation is largely derived from agricultural practices that have been carried out in recent decades, which has forced a present in which, "with the usual practices, there are very few sustainable diets" (Dernini, 2022: 21).

The relationship between human beings and nature has always been complex and marked by constant attempts of domination by men over an environment that was, at the same time, a source of wonder and fear. Over the centuries, the evolution of technology allowed increasing defiance against the territory and the imposition of decisions by humans, always within the coordinates of finite growth, where progress spread in a unilinear way, as an advance of the unstoppable submission of natural systems to the positivist will of human beings. It is evident that man shapes the landscape with his activity and, within this generality, in a fundamental way, with the development of the primary sector where agriculture and livestock are grouped and gives rise to the basic productive system of any human settlement. The incompatibility between a growth imagined as infinite in physical reality, and the planet Earth, which is finite, became apparent at the time when humans were not only able to destroy it, but also not able to project or maintain its indefinite exploitation.

Devouring nature without feeding people

As mentioned above, humans maintained a peculiar relationship between the use and enjoyment of the natural environment for centuries, in a balanced way, until the dominance of science and, with it, nature. In

---

[3] It should be specified that, even though anthropology always paid special attention to feeding behaviors and is the discipline from which this work is approached, it is the philosophy that raises at the ontological level the foundations of ecology and, in this sense, as Lourenço Fondevila says, "the proximity of the thought of Heidegger on *modern excess* and the despoilment of the land with the ecological thought (sometimes even with romantic-anarcho-ecological verbalism) is manifest." (1991: 605).

relation to agricultural activities,[4] around the middle of the 20th century, scientific and technical advances allowed humans to increase the productivity of the land with the introduction of varieties of high-performance genetically modified crops, known as the Green Revolution. This international denomination, where, as pointed out by different authors, the use of military-type language that focuses on its geopolitical character is significant, came about in the sixties after the development of experimentation programs on hybrid wheat seeds and corn in Mexico in the forties, although it later spread to other countries. The Green Revolution consisted of a spectacular increase in the production of basic foods for human consumption, particularly cereals (wheat, rice, and corn), through the use of HYV (high-yielding variety) seeds. In the beginning, the Green Revolution was presented as the silver bullet for problems as serious as malnutrition and famine, which affected a considerable part of the world population and, theoretically, could be eradicated through the massive production that was going to be achieved with the practices of modern agriculture, even as population growth continued to increase.[5] This was the case for farms because genetic selection also occurs in livestock. The general use of growth hormones was introduced, and the traditional feeding formulas were varied,[6] entering into practices described as modern that were presented as a definitive solution, without stressing the extreme

[4] We refer to agricultural activity as it is understood that it is, as a productive pole, one of the ends of the food chain that has an opposite pole, the consumption of the products by the consumers. In any case, food consumption would be followed by the fifth stage of those established by Goody (1995), which refer to the treatment of waste being the stage of collection or waste.

[5] In his work on the origins of the Green Revolution, Wilson Picado stresses how the "legitimacy of the battle against hunger also found religious and, to a certain extent, messianic arguments" because, "for Freeman, the extension of the North American agricultural technology and knowledge across the world was a process that was legitimized not only by a reason for obvious technical superiority but also by a moral obligation on the part of American governments" (Picado, 2008: 48). It should be noted that Orville L. Freeman was Secretary of Agriculture in the United States of Kennedy and Johnson and the author of *World without hunger*.

[6] In certain situations, such as the recent transport strike in Spain, the extreme weakness of a production system exposed to the margins of sustainability parameters is evident. The news gathered by the mass media on the risk of death by starvation for livestock can only be explained because in the transfer from peasantry to agricultural business, the circular pace of traditional production was broken. The model of the production unit that matches the family unit, where agricultural and livestock activity is oriented toward selfconsumption or toward a short-range distribution chain through local markets, fully disappeared. This reality, characteristic of areas such as Galicia not so long ago, gave way, because of modern times, to a concentrated production that can only be maintained within international circuits of heavy traffic of goods in such a way that animals raised in stables are fed with fodder and feed produced thousands of miles away in the same way that their derivatives are also consumed outside the territory.

dependency on external aggregates. This means the indispensable use of products of chemical origin in two variants: either as manure and specific fertilizers or as pest controllers and weed killers. The fact that the now so-called conventional agriculture would strongly depend on irrigation systems was also not perceived or exposed as a problem. However, without them, it would not have been possible to obtain the much sought-after results of increased production, when it is well known that water was and is a scarce commodity on the planet. Of course, they did not refer, in the narrative created at that time, to any future consequences in terms of food security other than an obvious improvement in the quantity of food that could be produced. This intended modernization, however, proved to be another tool of neocolonialism presented, undisguisedly, under the argument of the authority of scientific knowledge. Thus we observe how "the book by Theodore Schultz, a North American author known as one of the ideologists of the Green Revolution, *Transforming Traditional Agriculture*, emphasized that the agronomist was a person who would civilize the barefoot, the barbarian who was in close contact with nature, but subjected to it" (Ceccon, 2008: 23–24).

In reality, the general idea of Green Revolution was to modify the environment to create the required optimal conditions so that the seeds could reach their full potential. To our understanding, a key factor in the triumph of this proposal all over the world was that it was presented as modern, confronting the delays of traditional practices.[7] Thus the populations of numerous local communities were subjected to constant pressure from discourses that, from a global perspective, were reminding them, or rather, inoculating them with the idea that to be better or to advance, they had to get on the bandwagon of modernization. Whether this improvement went through a spectacular quantitative growth that would justify the abandonment of traditional practices was linked to the selection of the most optimal seed varieties according to the areas and characteristics of the land, carried out for centuries empirically by farmers, and the dismantling of the cultural diversity of each place, which was neither valued nor questioned: The paradigm was presented as unquestionable and sold as absolute.

---

[7] It should be borne in mind that the peculiarities of each area were fundamental to traditional agricultural knowledge and practices, reflecting a good biological and cultural adaptation to the environment, as "in an evolutionary sense, the process of adaptation to the environment involves the selection of those genetic and cultural characteristics of the species" (Hardesty, 1977: 44).

Traditional farmers worldwide were standardized and homogenized under the totalitarian coating of progress and started a process of acculturation in parallel to the loss of natural biodiversity. The "new" peasants had to adapt to new agricultural practices and new high-performance seeds to achieve optimum production levels so that they could be considered on an economic level and, therefore, on a social level. To be a modern farmer was linked with the use of high-performance seeds in combination with the required external aggregates (the chemical companies and laboratories were selling packs that included seeds, fertilizers, and pest controllers). This required modifying the environment to achieve the most favorable conditions so that these productive varieties (reduced in number) could reach their maximum fecundity, rejecting traditional knowledge and breaking the chain of generational transmission of agricultural knowledge.

The success of the Green Revolution regarding its dissemination worldwide was directly related to other phenomena: internationalization and globalization, which the societies of the entire world were immersed in during the last decades of the 20th century. In agriculture, as well as other fields, unitary solutions arose to be applied in different places, with different characteristics and conditions, standardizing what until then had been local or regional peculiarities in terms of production and technical knowledge of agricultural practices. These practices, at the same time, confirmed the idiosyncrasies of each town, affirming both the sharing of some common practices and the differences with respect to others. The established objective was the economic fallacy of the global market, where, supposedly, the modern producers were going to sell their consumer goods, shaped as *El Dorado* to be discovered, which was pursued and longed for, and where the laws of operation were not subject to discussion, it was only necessary to enter the system.

Eighty years after the promised manna from heaven, the current situation does not meet those expectations created around intensive agriculture as advocated by the Green Revolution. The problems associated with modern agriculture emerge from several fields, starting with agricultural sciences, which question not only the foundations, from a scientific point of view, of the so-called Green Revolution, but also the annulment of the negative assessments or questions that this project had from the beginning.[8]

Now we are in a moment where agriculture worldwide is in permanent crisis because the relentless growth is not finding ways to sustain in

---

[8] Doctor Norman Borlaug, main promoter of the Green Revolution, already acknowledged and brought to light several "mistakes" regarding the use of fertilizers when in 1970 he won the Nobel Peace Prize acknowledging his achievements (Sarandón, 2002: 37).

the long term. It has also been proven that the modern intensive agriculture that displaced traditional techniques produces much lower yields than usual in the absence or failure of an external aggregate. Along with this technical issue, at the other end of the production chain, that is, from the consumers of agricultural and livestock products, more and more questions are raised about the quality of the products of conventional agriculture compared with those of traditional agriculture. Lastly, beyond the debate on how to maintain the quantitative high level of production in conventional agriculture or whether the final quality of the products obtained through these practices is superior or inferior than the quality of products obtained through ecological agriculture, where traditional knowledge may find a second chance, the debate that is approaching in the current society is whether one type or another of practices, conventional or ecological, are more or less respectful of the environment, whether they help us to think of the human being as a resident of the planet to preserve and pass on to future generations or simply to use them and make them profitable in economic terms that will be of little use when natural resources are neglected. In this situation, the moral dispute, the current challenge of agricultural production is in the creation of a sustainable alternative that allows the tackling of a situation where "the symptoms are so clear, the problems are so serious, the consequences so worrying that even those who were emphatically denying them, have started to acknowledge them, the evidences are overwhelming" (Sarandón, 2020: 14).

From global to local: food consumption in the city of Lugo

Lugo is the capital of one of the four provinces that constitute Galicia, an autonomous community in the northwest of the Iberian Peninsula, which is therefore a part of the Spanish state. With just 97,613 inhabitants (https://www.ine.es/jaxiT3/Datos.htm?t = 2880) Lugo is an urban center where the population works, mostly, in the tertiary sector with a strong presence in the administration and services. However, its cultural route appears infiltrated by the rural and, linked to this, by the productive primary sector. Suffice it to say, as an allegory, that in the tour of its most outstanding architectural monument, the Roman wall built between the third and fourth centuries of the Roman era and registered as a World Heritage Site by the UNESCO in the year 2000, it is impossible not to glimpse the green color of nature on the horizon. During the slightly more than two kilometers of the monumental perimeter, it is not possible to lose sight of the line of land that physically marks the cultural border, and the idiosyncrasies of Lugo remain present both inside and outside the Roman wall.

This bond with agriculture is one of the differential elements for the province within the Galician reality, already connoted in a particular way by circumstances. If agrarian Galicia is a common place in the whole of the Spanish state, the province of Lugo seems to be (and the data coincides with this) a stronghold for the sociocultural remnant that represents the peasantry.[9] A safeguard of rurality that, even when dismantled in economic terms, remains in certain symbolic and pragmatic associations with regard to food, manifested on multiple occasions.[10] In this line of thought, the echoes of the strong presence of the primary sector in the province carry to the city of Lugo an echo that identifies it in sociologic parameters as a village within the Galician urban fabric, and it is not unusual to say that Lugo is not a small town but rather a large village.[11]

It is understandable that observing the consumption patterns of the population of Lugo has created notable interest. This interest is due to the fact that in the city of Lugo, the overwhelming economic reality has forced the footprint of a Green Revolution connected with the European Economic Community (EEC) accession in 1986, now European Union, and in this transformation has created the marks of nutrition and gastronomy based on tradition, which is expected to have a much more sustainable profile. There are two reasons to observe these consumption patterns: on the one hand it would be to check how the maximum proximity, both spatial and temporal, as characteristics of this territory, interferes

---

[9] All in all, in Galicia, "the first fact that must be highlighted (...) is the intense decrease that the relative weight of the agrarian sector has experienced, both in terms of production and especially in employment. In this respect, over the last thirty years it has culminated in what has been one of the fundamental transformations of Galician society since the middle of the 20th century: the late but abrupt deagrarianization"(López-Iglesias, 2019: 3).

[10] The slogan "E para comer Lugo" ("And to eat, Lugo"), originated in an advertising campaign by the Ministry of Information and Tourism in 1962, is currently maintained as the city's identifier (See https://www.lavozdegalicia.es/noticia/lugo/portomarin/2021/04/21/celebrada-primera-cata-comer-lugo-camino-portomarin/0003161902409354811962.htm. *El Progreso*, 23/02/2022 o https://eparacomerlugo.com/).

[11] The result of this particular concept is the statement that everyone in the city knows each other. To go deeper into the subject, see Lamela Viera (1998).

between production and food consumption. On the other hand to consider the possibility of observing, on an accessible scale, the impact of global movements on local realities. In this specific context, how is the transition to a sustainable society produced, and what are the circuits through which organic food is transported? Do these organic products represent the total of the practices understood as sustainable?

Resuming the cultural perspective, the fact that the concept of feeding goes beyond the mere intake of food has been considered a major concern by anthropological studies. It is normal for research on dietary habits to contain this perspective of social and cultural order. This would explain, among other topics, the criteria of acceptance or rejection of a piece of food in a specific sociocultural environment, the symbolic values that the piece of food has or those values awarded to it, the appearance of new feeding habits or the continuity of others, etc. This does not restrict the inherent nutritional values or the analytical perspective. These values do not necessarily have to coincide with the first ones, as not all edible products are necessarily consumed. But, moreover, it is crucial to address any concept of environmental sustainability because the environment or the prospects made about it may be, in any case, alien to anthropogenic action.

As has been announced, it is necessary to note the peculiarity of this research — this research is circumscribed to a traditionally rural area such as the city of Lugo, where many agricultural and livestock resources, and therefore food resources, labeled as traditional, still subsist. In this sense, the presence of organic food coexists, as a cultural manifestation, with other symbolic systems that allude precisely to the rooting in the collective culture of certain values based on the rural origins of the vast majority of the population. The feeling of community cohesion and the attachment to tradition turn these foods into products that, necessarily, must have established a specific symbolic space that differentiates them from other areas that may be very similar, as in the case of organic foods.

First, a deeply rooted value in Galician culture is the reference to *da casa* (homemade), applicable for certain foods that can be found in somewhat alternative economic circuits, although completely traditional. This refers to consumption goods that arrive directly from the producers to the consumers, without any postharvest treatment and that can be found at traditional fairs, local markets, or certain minority locations. Some foods such as animals, usually small, but on some occasions also large cattle, various seasonal vegetables, eggs, etc., considered to have certain specific value, have this denomination *da casa*, referring at the same time to the physical and, above all, affective and symbolic proximity to a place connoted in Galician culture with inaugural and mythical values, such as the rural environment.[12] Usually consumers have some direct reference regarding the place of origin of the product and thus its intrinsic and, logically, subjective qualities. This reference does not necessarily have to be direct and personal, but is part of an absolutely traditional system of guarantees, which does not depend on labeling, quality seals, official guarantees of origin, etc.; on the contrary, its validity has to do with personal knowledge, with the direct knowledge of the people involved in the exchange, and, above all, with a previous knowledge of the production system and the place of elaboration with which the buyer tends to identify. Therefore, the prototypical consumer of this kind of perishable goods tends to be a person who is personally or familiarly related to the land and who maintains with it a series of affective bonds and symbolic values that allow them to obtain from the product not only a food but also a sign of identity.

These types of foods are not necessarily organic, according to the precise definition of the term.[13] They usually do not respond to these characteristics, as they come from an agriculture that is nonorganic but conventional in its production, with forms of conservation or processing that follow the traditional pattern. In reality most of these products are surplus or minority production for a very limited market, economically boosting the families producing them. However, these are goods that refer to a geographic origin, to a traditional type of elaboration, craftsmanship, to a specific type of conservation or consumption that are related to subjective estimations attributed to them collectively in which quality, food values, gastronomic and social symbolic values (fusion with a community that eats that food, a group cohesion factor, that is, identity),[14] those associated to health and even much less precise notions such as naturalness, the same that, for example, and in an inverse sense, tends to make

---

[12] The relevance of the concept "casa" (home) in Galician culture is evident when consulting all the anthropology carried out on this territory and is paramount to follow our line of argument, however, for a matter of space, we will only refer to Lisón Tolosana, who affirmed that "each house is (...) a social structure. And also a mental texture." (Lisón-Tolosana, 1978: 112).

[13] By organic food we mean food from organic agriculture, that is, from a set of agricultural techniques that exclude the use in agriculture and livestock farming of synthetic chemicals, such as fertilizers, pest controllers, antibiotics, etc., with the aim of preserving the environment, maintaining, or increasing the fertility of the soil, and providing food with all its natural properties.

[14] To this respect, Lorenzo Marianto et al., in their paper on gastronomy from Las Hurdes, which tackles the translation of the stigma of the hunger that marked this territory during the 20th century to the abundance with which it is identified today, affirm that "high quality thus maintains a strong relationship with the territory" (Mariano; Rivero; López-Lago; Conde-Caballero, 2020: 24).

transgenic foods unreliable compared with those considered natural, regardless of the objective contents of one or the other.

Together with this commercial and symbolic structure, even though it is still important for minorities because it represents the survival of their own cultural patterns, there is another type of new product, the so-called organic products, that answer certain specific conditions. For example, unlike the previous products, its origin does not have to be precise, known, or close in the sense that there is no personal, direct, or affective relationship between the consumer and the place of origin of the product. For this reason, perhaps the traditional values of the identical could not be attributed to them. Although it is also true that there is a system of advertising of this type of product, as well as others that circulate in parallel and do not respond to the same definition,[15] that aims to adopt these same values related to the next, to the own, and therefore to the natural and the good.

In fact, if we analyze the marketing resources used to present organic products to the consumer, we observe a majority preference for a neutral iconographic aesthetic, in which the range of ocher and green tones predominate, clearly linked to the land and with packaging that we can basically differentiate into two types according to their aesthetics and the product contained: the ductile and the rigid. The first one can be made of plastic, paper, or textiles and have a format that resembles a sack, both in its shape and its weaving system, sometimes sewn with straw or raffia. This type of packaging alludes directly to an evocation of the past when products were sold in bulk,[16] although it is now used as a system of presentation and individual packaging. The latter can be made of clay,

---

[15] The Ministry of Agriculture, Fisheries and Food itself provides information under the slogan *De lo nuestro, lo mejor* (Of our products, the best) on differentiated quality foods, including products with "Protected Designation of Origin," "Protected Geographical Indication," "Guaranteed Traditional Specialties" and "Organic Production" products, unified under a single seal of guarantee. The graphic image of this is the result of a contest for its modification in 2016, which was won by the designer David Delfín. Since then, the logo used in the first image has been modified by the second image.

[16] This system of retail sales is still being used as we will see in some stores specialized in the sale of organic fair trade products.

preferably for dairy products, wood or glass, the latter being the most abundant, since they are used in packaged products with a longer expiration date. The use of glass responds, in some cases, to a reinforcement of the ecological role of the producers, since they are returnable containers, whose reuse is intended to conserve and protect the environment by reducing the production of nonorganic waste that cannot be returned to the land for organic restructuring. But also, these containers, which in most cases are not picked up again by the stores, also offer other associated characteristics that are equally important from an advertising point of view to be chosen. We refer to the evidence that the glasses used are transparent,[17] that is to say, it does not hide the content but rather presents it clearly, reinforcing the underlying idea that there is nothing to hide, that what is presented there is the best and that by its very presence it convinces the consumer. Another issue we should mention is the slogans, legends, or labels that identify the different products and, within this point, we must highlight two differentiated traits. First, the type of typographies used, which are calligraphy imitating old typographies, with a clear preference for letters inclined to the right, and calligraphy of children's typography, which shares with the first one its unequivocal tendency toward the rustic and the past. Both possibilities reinforce, each in its own way, the values of the own, the natural, and the good. Thus, while the first type of typography puts us in relation to the truth that is being transmitted to us by someone enlightened and even erudite, who knew how to write in our closest collective past? The second uses the resource of the innocence of children highlighted by the evocation of their own individual childhood, also as a guarantee of the truth. It is very important to stress that the use of handwritten, calligraphic typographies, even though created and reproduced in an industrial manner, suggests a spokesperson who invites a supposed dialogue with the consumer: someone wrote that for them, for "me," something that also invites exclusivity, which is also closely related to a sector of current consumers of organic products.

The second characteristic regarding the signs or labels used is the type of message they transmit. They make a clear and explicit reference to the place of origin with total precision and accuracy, identifying that food with a very specific space with which the consumer can connect, even if they don't

---

[17] This quality of formal and subconscious transparency is also present in the case of the plastic packaged products, which allow sight of the internal content without "hiding."

usually have any personal relationship with the place. This localization may be accompanied by some other referential data regarding the surroundings or the landscape where this product is produced or manufactured. This provides additional information for the consumer, who can acquire, when buying this product, a natural experience or contact with nature, for example with *Mel dos montes do Courel* (Honey from the Mountains of O Courel). The simplicity of the labels should also be noted, which make direct reference to the type of base product being worked with. A good example, among others, would be Arqueixal, a company traditionally dedicated to the manufacture of cheese and, more recently, to yogurt, or *Ganderia Quintián*, which sells milk[18] directly to the consumers.

Precisely, the symbolic qualities that can be associated with organically produced foods coincide, in part, with the qualities established in the case of the traditional, homemade products mentioned above. If this does not occur at all, it is because the public for whom the homemade (*da casa*) food is intended and the public for whom the organic food is intended, are not the same. In the case of the former, the main consumers are people with family ties to the land and the cultural and social system that the foods come from. Buyers of organic food, in addition to economic considerations about the price of products that are considered more affordable, do not tend to make these economic considerations an absolute value. Likewise, they are usually people that lack a direct relationship from a personal point of view and, undoubtedly, maintain a value system on feeding that belongs to tradition. This system has to do with the quality and the naturalness of the foods of the past, but that go together with other types of foods of new creation that are related to ecology, a different socioeconomic awareness and a more technical precision in the definition of the concept of natural food. Thus, a consumer of organic products does not have to value traditional manufactured foods as positive because it can be considered as a deterioration of the traditional forms due to the emergence of new forms of processing or preservation, by the use of certain chemicals that are not considered optimal, etc. At the same time, for a consumer of homemade foods, organic foods can lack the identity signals and the symbolic values that homemade food has.

Therefore, these are two different concepts of the natural and the traditional. In essence, the buyers of homemade food usually have a less purist attitude — they admit the possibility of using methods and products

---

[18] We are not going to establish an exhaustive list of the many companies that are representative of this case in the city of Lugo, but their volume has increased exponentially in the last decade.

that are not totally traditional in the cultivation of these foods, although not in their preparation, but it is circumvented or concealed by the symbolic values, the origin and the cost. For the consumers of organic foods, however, the main choice is the production system that cares about the health and the sustainability of the agrosystem, the biodiversity, and the biological cycles of soil. According to the definition of the *Codex Alimentarius* of the FAO/WHO, it should use agronomic, biological, and mechanical methods rather than chemical synthesis, resulting in food free from residues and pollution and not originating from genetic alterations. Only as a consequence of these practices would the foods be considered to have values such as health, culinary quality, or naturalness, etc.[19]

In view of the above circumstances, it is obvious that there is a coincidence on the basis between these two types of products, because both of them go back, as an ideal reference model, to traditionally produced foods prior to the contemporary agricultural revolution. It is evident that the historical or social circumstances do not allow, and perhaps do not make viable, the restoration of this production system; however, we try to make an approximation that is as faithful as possible to this model, in each case according to the particular considerations of the tradition. This common model affects the definition of the type of objects of consumption, as has already been said: it is about goods elaborated according to a series of criteria that result in food that is unaltered, supposedly richer, and more nutritious and, therefore, better. It is essential to bear in mind that, from the consumers' point of view, this assessment is based on what Alexandre Meybeck and Vincent Gitz define: "We propose to denominate these characteristics 'quality' attributes, and note that they are all 'credential attributes' which cannot be tested by consumers and for which they have to rely on information" (Meybeck & Gitz, 2014: 181). The main difference is that in the case of organic products this information should be stated, at least on a theoretical level, on the mandatory labeling, but in the case of artisan products the channel through which it is transmitted may be oral, in some cases, but above all it is produced by contextual parameters that affect the confidence that have the homemade (*da casa*) element as a strategic focal point.

---

[19] The European legislation on organic agrarian production is based on the EEC Regulation 2092/91. In Spain, since 1993, the legal powers on the subject have been transferred to the autonomous communities, which are those that define the control bodies which, in turn, facilitate the identification of products of this type of production, although there is also, since 2000, the possibility of using a European organic label.

One offer that expands against another that contracts?

Another distinguishing factor between organic products and those of traditional production that continue to be maintained today has to do with the commercial and economic entity that they have today. While organic foods are experiencing constant expansion,[20] traditionally produced or homemade foods, even the gastronomy associated with them, can be considered to be contracting, parallel to the substitution of traditional market and production systems by other types of manufactured products. The systems of distribution and sale are related to this issue. In the case of the foods that we call homemade, these are perishable and fresh products in general, not subject to any type of packaging process or conservation mechanism. Their distribution networks, as already mentioned, are limited to traditional circuits such as markets, fairs, gastronomic festivals celebrating traditional products, and, in some cases, commercial and gastronomic places of small volume that distribute them in a parallel and independent way from the conventionally established networks, in some cases even outside the sanitary controls used for these types of products.

On the other hand, organic foods are marketed in other types of markets in specialized places such as in pharmacies or with herbalists or dietary product stores. In this case, the way to reach these foods arises as a complement to its main activity related to alternative or natural medicine and the supply of products that help to maintain a controlled diet that favors the maintenance of the diet either for aesthetic or medical reasons; through the link with healthy and natural products, the leap to organic products is immediate and is given by a simple logic of association, often motivated by requests from the customers themselves. Regarding this form of consumption in the city of Lugo, it is limited to organic products packaged and preserved.

A second market that is also specialized but more sophisticated is that established through a collective initiative with groups of consumers of a cooperative

---

[20] In fact, as reported by the SEAE on its website, in 2020 Spain was the first European country in surface and organic production and the third in the world, behind Australia and Argentina, and in a position to meet in 2030 the target set in the European Green Pact to allocate 25% of agricultural land to organic production (https://www.agroecologia.net/estadisticas-produccion-ecologica-espana-2020/). On the other hand, the Ministry of Agriculture, Fisheries, and Food informs that in the period 2015–20, the growth of spending on organic products was 67.42%, an increase of 154.07% if we take a longer reference period (2012–20). (https://www.mapa.gob.es/es/prensa/ultimas-noticias/el-gasto-en-productos-ecol%C3%B3gicos-en-espa%C3%B1a-se-incrementa-un-7-con-respecto-al-a%C3%B1o-anterior-/tcm:30-583763).

or associative character and with the specific purpose of providing its members with a series of products that they demand and that, at a certain moment, were difficult to find in the traditional market even though they did not correspond 100% to homemade products. In this type of establishments it is much more common to find fresh and perishable goods that are usually complemented with others from so-called fair trade, even though this type of store does not always have products from organic farming. This section would include O Bandullo Ecolóxico, a consumers association where they give "priority to fresh and unpackaged, seasonal, local, first necessity products that can be bought directly from the producers" (Vázquez Castro, 2016: 3). In the case of this consumers association, it should be highlighted that it not only enables access to organic products but also to the so-called artisan product, with which the lines between the different categories may be overlapping and affect the complexity that we have been advancing in our analysis.[21] This symbiosis is complete in the *Mercado da Terra* (Land market), a trading network promoted by the Bandullo Ecolóxico (in original Galician language) where, once a week in the municipal market, but in an afternoon schedule, not to coincide with the usual itinerant activity carried out in the morning schedule, organic and artisan producers sell their products directly to the general public.

In the same line would be the stores A Cova do Raposo, with a transfer not exempt from controversy and mobilizations, which occupied the place of A Cova da Terra as well as Bico de Grao or Nai Terra. All these stores sell organic packaged and fresh products next to other products that can be food, or not, coming from fair trade. Although access is free, there is the figure of the member with access to special discounts thanks to belonging to the association, in a gesture to promote consumer awareness of the problems not only of a nutritional or ecological nature but also focusing on other social inequalities. Within this task of collecting and promoting a rational, balanced, and responsible consumption, they also offer talks and conferences on diverse issues but all of them in the line of people aware of issues that, generally, revolve around the exploitation of natural and human resources.[22]

---

[21] In the "artisan" category, the ethics of fair trade are also included. In this case, on a local scale, it is linked to the sustainability of food consumption in short or zero kilometer circuits.

[22] The approach to organic products through a deep reflection on the current situation of the world is also exposed from an analysis of the emergence of ecological movements that took place in the most developed places in the world, that are at the same time the most responsible for the situation created by their consumerist habits. In fact, "the economists and environmentalists from developing nations complain that the negotiations agenda on environmental issues is almost exclusively decided in the North. They affirm that, with the excuse of saving the planet, the industrialized world exerts a new type of domination, which is "ecoimperialism" (Sidhva, 2001: 41).

The third possibility is in large commercial areas that in the last decade tended to establish sections specifically dedicated to this type of product. Going from having a reduced offer and embedded in an ordinary way in the regular lines of sale to occupying a significant location, both in terms of surface area and range of products available, to the point that some supermarket own brands have added a bio section. It is in these large stores where the identification by law of the foods coming from organic farming with a distinctive logo is important. Taking into account the variability of information available to consumers who make their purchases in these stores and the presence of all types of qualities within the products, as well as the lack of personal links that guarantee the purchases in this type of store where precisely the anonymity in transactions is fostered, it is easy to understand that an identifying mark is needed so that it will offer the consumers the certainty that the products they are buying come from organic farming.[23]

The fourth possibility to purchase products from organic farming in the city of Lugo is based on a very relevant association that links these goods with the so-called *delicatessen* or gourmet, both for the space they occupy, the aesthetics they employ, and the price they tend to have. This type of store is where it is possible to appreciate that organic foods converge with other high-end products and are configured as indicators of social differentiation

---

[23] The debate about the scope of the information conveyed to the consumer in the labeling and the possible confusion it causes occupies a prominent place in studies of food consumption. As an example, we will cite that in 2019, the government ordered to remove the word Galicia from the Regulatory Council image to avoid "consumer confusion and collisions with designations of origin" (La Voz de Galicia, January 20, 2019. Available at https://www.lavozdegalicia.es/noticia/somosmar/2019/01/20/gobierno-ordena-retirar-galicia-sello-agricultura-ecologica/0003_201901G20P36992.htm). Currently, the organic certification uses the following labeling:

Name: EU Organic Bio Geographical scope: European Union
Products: All EU food products prepackaged, produced, and sold as organic within the EU

Name: Organic farming seal from the autonomous communities geographical scope: Spain
Products: Agriculture, livestock, and fisheries

that, for the consumers, guarantee a higher status in direct relationship with the designations of origin. Again, this is the case of a physical location that works as a link between the land and the buyers and with the quality of the products that are offered. In these stores they sell the best, the most natural, and the most exquisite for the palate but also for the mind; here, the delight associated with consumption reaches its highest levels.

To put an end to this section, we must mention another two sale points for the organic products that are flourishing in recent times. The first one is not linked to any specific place, these are more or less traditional fairs or gastronomic celebrations, which start to have specific stands for organic food. Of course, in these places more than in any other type of store, the symbiosis between the traditional and the organic is complete, although it should not lead to confusion. Consumers of organic products that buy them at a fair, often for the first time, purchase the products in an environment evocative of rural areas, affectively close by, with those added values that homemade products have. They understand these products not just as a nutritious substance but as a possibility of responsible maintenance of the natural environment and of development of the rural area that is so close and, at the same time, so far away from the urban profile of most of the people that attend this type of fair and events. Precisely in this context, in the confluence between a healthy and a responsible consumption of basic needs and the empowerment of rural areas, in this case, from Lugo, the organic products open a gap in today's consumption habits, which can be, from the economic and also the social point of view, a useful response to the inevitable globalization.

Regarding the second point of sale, it is made up of different food stores that have spread across the city in the last fifteen years and, in the case of the most privileged areas in socioeconomic terms, have combined their aesthetic of a neighborhood store with an offer of products where you can find organic, artisan, and conventional products wrapped in a double layer in which two attributes are decisive: quality and belonging. In this context in which the denomination itself comes into play in almost all the cases, with the inclusion of the own name, surname, or nickname of the owner within the commercial name, this would imply that all the products are ours in as much as they answer, at least from the narrative point of view, to the definition of "trusted" products, trust earned by the homemade food (*da casa*) in other times and spaces (for example, the local market). One particular case within this type of store, because of its specificity in the offer of nonperishable products and without packaging, would be O Graneiro de Amelia (Ograneirodeamelia.com).

A final section should be established with the appearance of a new modality of food distribution that is related to establishments that sell ready-made food to take away. In this category in Lugo we can find, since 2017, the proposal of Ecoxantar (https://www.facebook.com/Ecoxantar/).

Conclusions: the future was yesterday

We have seen how in small contexts, highly intervened by geopolitics and global economy, the sustainability guidelines that permeate today's society have made their presence felt at an early stage that we can place in the change of century. These guidelines have offered the consumers organic products that are perfectly identified and distributed through commercialization networks such as specialized shops closely linked to health, just when the benefits of dietetics and nutrition started to spread, as well as in large supermarkets. However, in a parallel way and mixed with the support of tradition, this limited area of organic labeling was provided by a whole series of traditional practices that, apart from the formal processes of recognition, maintained the confidence of consumers, even though it was sometimes false. The aforementioned practices made use of the complexity of a cultural system that, while accepting and putting into practice the foreign impositions in the food production phase, maintained its traces in the types of conservation, in the distribution networks, in the preparation methods, or in the consumption times under the homemade label.

In the awareness that is required from the immediate present of the consumers, for the sake of sustainability that makes the future viable, it is paradoxical, and we want to emphasize this, that in a local reality of eminently rural character the global guidelines now place the ethical responsibility on them to maintain the place from which the socioeconomic dynamics have expelled them. In this perception that the food chain can be influenced by the control capacity of consumers insofar as the system, marked from an economic perspective, needs them to sustain itself, it is undeniable that food, codified as a commodity, constitutes a priority field of work to achieve sustainability. In a reality in which most of the population does not produce food, but consumes it without a previous relationship, the possible choice becomes a tool of pressure but also a responsibility.

Well, in a context such as the one studied here, the current circumstances point to the fact that it is the children or grandchildren of the farmers who have been charged with abandoning or transforming traditional agricultural production to adhere to the longed-for modernity that would drag them to social ascent who have found in agroecology an ember that connects a certain remnant of the past with the dynamics of a sustainability "raised from the need to integrate three dimensions (. . .) socially equitable and fair; socially equitable and fair; economically viable and efficient; and ecologically safe and rational" (Medina Rey, 2015: 16).

That option for the future includes, of course, a gastronomy that incorporates organic food, but also all the baggage of what, in a collective way, we ourselves were when our relationship with nature, instead of being a hardship, ensured our future in terms of food and environment. The heirs of a community that was mostly settled in a successfully productive rural area and that was dismantled now become enablers, from another side of the chain, of a future in which food can become a catalyst for their own right to be sustainable while they return to their own culture.

References

Aguilar, P. (2014). Cultura y alimentación. aspectos fundamentales para una visión comprensiva de la alimentación humana. *Anales de Antropología, 48*(1), 11−31.

Ceccon, E. (2008). *La revolución verde tragedia en dos actos. Ciencias* (Vol. 1, pp. 21−29). México: UNAM, Núm. 91, julio-septiembre, 2008.

Contreras, J., & Gracia, M. (2005). *Alimentación y cultura: perspectivas antropológicas.* Barcelona: Ariel.

De Garine, I. L. (1976). Food, Tradition and Prestige. In D. N. Walcher, N. Kretchmer y H. L.Barnett (eds.), Food, Man and Society, Nueva York/Londres, Plenum Press, pp. 150-171.

Dernini, S. (2022). Dietas sostenibles en sistemas alimentarios sostenibles: retos para el presente y para el futuro. In Medina, Mariano, Conde, & Aguilar (Eds.), *Consumo alimentario y sostenibilidad ¿Hacia una sociedad sostenible?* (pp. 19−42). Barcelona: MRA ediciones.

Goody, J. (1995). *Cocina, cuisine y clase. Estudio de sociología comparada.* Barcelona: Gedisa (Colección Antropología).

Hardesty, D. L. (1977). *Antropología Ecológica.* Barcelona: Ediciones Bellatera.

Harris, M. (1990). *Bueno para comer.* Madrid: Alianza.

Lamela Viera, M. C. (1998). *La cultura de lo cotidiano: estudio sociocultural de la ciudad de Lugo.* Madrid: Akal.

Lévi-Strauss, C. (1972). *El pensamiento salvaje.* México: Fondo de Cultura Económica.

Lisón-Tolosana, C. (1978). *Ensayos de Antropología Social.* Madrid: Akal Universitaria.

López-Iglesias, E. (2019). O sector agrario e agroalimentario en Galicia; balance das transformacións desde a integración europea, 1986−2016. *Revista Galega de Economía, 28*(3), 1−20.

Eating and thinking at the same time

Medina Rey, J. M. (2015). El derecho humano a la alimentacion en los tiempos de la sostenibilidad. *Revista Ambienta, 113,* 2—21.

Meybeck, A., & Gitz, V. (2014). Signs to choose: voluntary standards and ecolabels as information tools for consumers. In Meybeck, & Redfern (Eds.), *Voluntary standards for sustainable food systems: Challenges and opportunities* (pp. 171—186). Roma: FAO.

Picado, W. (2008). Ciencia y geopolítica en los orígenes de la Revolución Verde. *Revista de Ciencias Ambientas (Tropical Journal of Environmental Sciences), 36*(2), 46—56.

Sarandón, S. J. (2002). *Agroecología: El camino hacia una agricultura sustentable.* La Plata, Buenos Aires: Ediciones Científicas Americanas.

Sarandón, S. J. (2020). Agrobiodiversidad, su rol en una agricultura sustentable. In S. Sarandón (Ed.), *Biodiversidad, agroecología y agricultura sustentable* (pp. 13—36). La Plata (Buenos Aires): Edulp.

Sidhva, S. (2001). Ecoloxía, o imperialismo disfrázase de verde. In *O Correo da Unesco* (pp. 41—42), París.

Vázquez Castro, M. D. (2016). *Funcionamento e evolución dunha asociación de consumidores: o Bandullo Ecolóxico" in* Actas del XII Congreso SEAE. *Leguminosas: Clave en la gestión de los agrosistemas y la alimentación ecológica.* Lugo: SEAE.

Further reading

Ab Karim, S., & Chi, C. G.-Q. (2010). Culinary tourism as a destination attraction: An empirical examination of destinations' food image. *Journal of Hospitality Marketing & Management, 19*(6), 531—555.

Achmad, Z. A., Ida, R., & Mustain, M. (2020). A virtual ethnography study: The role of cultural radios in campursari music proliferation in East Java. *ETNOSIA: Jurnal Etnografi Indonesia, 5*(2), 221—237.

Albuquerque, C. R. D., Mundet, L., & Aulet, S. (2019). The role of a high-quality restaurant in stimulating the creation and development of gastronomy tourism. *International Journal of Hospitality Management, 83,* 220—228.

Barroso, F. (1985). Los Moros y Sus Leyendas En La Serranía de Las Jurdes. *Revista de Folklore, 50.*

Batat, W. (2021). The role of luxury gastronomy in culinary tourism: An ethnographic study of Michelin-starred restaurants in France. *International Journal of Tourism Research, 23*(2), 150—163.

Buttriss, J., & Riley, H. (2013). Sustainable diets: Harnessing the nutrition agenda.". *Food Chemistry, 140*(3), 402—407. Available from https://doi.org/10.1016/j.foodchem.2013.01.083.

Castillo-Manzano, J. I., Castro-Nuño, M., Lopez-Valpuesta, L., & Zarzoso, Á. (2020). Quality versus quantity: An assessment of the impact of Michelin-starred restaurants on tourism in Spain. *Tourism Economics.* Available from https://doi.org/10.1177/1354816620917482.

Catani, M. (1998). Las Hurdes Desde Dentro y Desde Fuera. In *Crónica Del II Congreso de Hurdanos y Hurdanófilos* (pp. 119—136).

Catani, M. (1999). Las Hurdes Como Imagen de Una Sociedad Local En Transformación. *Revista de Estudios Extremeños, 55*(2), 605—632.

Conde-Caballero, D., Rivero-Jimenez, B., & Mariano-Juarez, L. (2021). Memories of hunger, continuities, and food choices: An ethnography of the elderly in extremadura (Spain). *Appetite, 164.* Available from https://doi.org/10.1016/j.appet.2021.105267.

Conde-Caballero, D., Rivero-Jiménez, B., & Mariano Juárez, L. (2022). De La Abundancia a La Exquisitez Gastronómica. Una Etnografía de Las Nuevas Oportunidades Para La Atracción Del Turismo En Las Hurdes (Cáceres). *PASOS. Revista de Turismo y Patrimonio Cultural, 20*(2), 453—464.

Dacin, M. T., & Dacin, P. A. (2008). Traditions as institutionalized practice: Implications for deinstitutionalization. *The Sage Handbook of Organizational Institutionalism, 327*, 352.

De Jesús-Contreras, D., & Medina, F. X. (2021). Turismo Gastronómico, Productos Agroalimentarios Típicos y Denominaciones de Origen. Posibilidades y Expectativas de Desarrollo En México. *Journal of Tourism and Heritage Research, 4*(1), 343—363.

Emans, B. (2019). *Interviewing: Theory, techniques and training.* Groningen/Houten: Routledge.

Estalella, A., & Ardévol, E. (2011). E-Research: Desafíos y Oportunidades Para Las Ciencias Sociales. *Convergencia, 18*(55), 87—111.

Estalella Fernández, A. (2018). Etnografía de Lo Digital: Remediaciones y Recursividad Del Método Antropológico. *AIBR: Revista de Antropología Iberoamericana, 13*(1), 45—68.

Fernández-Poyatos, M. D., Aguirregoitia, A., & Bringas, N. L. (2019). La Cocina de Producto: Seña de Identidad y Recurso de Comunicación En La Alta Restauración En España. *Revista Latina de Comunicación Social, 74*, 873—896.

Fischler, C. (1995). *El (H)omnívoro.* Barcelona: Anagrama.

Gutiérrez-García, L., Labrador-Moreno, J., Blanco-Salas, J., Javier Monago-Lozano, F., & Ruiz-Téllez, T. (2020). Food identities, biocultural knowledge and gender differences in the protected area 'Sierra Grande de Hornachos' (Extremadura, Spain). *International Journal of Environmental Research and Public Health, 17*(7). Available from https://doi.org/10.3390/ijerph17072283.

Hammersley, M., & Atkinson, P. (1994). *Etnografía. Métodos de Investigación.* Barcelona: Paidós.

Harris, J. E., Gleason, P. M., Sheean, P. M., Boushey, C., Beto, J. A., & Bruemmer, B. (2009). An introduction to qualitative research for food and nutrition professionals. *Journal of the American Dietetic Association, 109*(1), 80—90.

Hine, C. (2011). *Etnografía Virtual.* Barcelona: Editorial UOC.

Holbrook, M. B., & Hirschman, E. C. (1982). The experiential aspects of consumption: Consumer fantasies, feelings, and fun. *Journal of Consumer Research, 9*(2), 132—140.

Jiménez Beltrán, J., López-Guzmán, T., & Santa-Cruz, F. G. (2016). Gastronomy and tourism: Profile and motivation of international tourism in the city of Córdoba, Spain. *Journal of Culinary Science & Technology, 14*(4), 347—362.

Kivela, J., & Crotts, J. C. (2006). Tourism and gastronomy: Gastronomy's influence on how tourists experience a destination. *Journal of Hospitality & Tourism Research, 30*(3), 354—377.

Kvale, S., & Brinkmann, S. (2015). *Interviews: Learning the craft of qualitative research interviewing.* London: SAGE Publications Inc.

Legendre, M. (1927). *Las Jurdes: Étude de Géographie Humaine.* Feret.

Lourenço Fondevilla, B. (1991). Heidegger: ecoloxía e o sentido do sagrado. *Grial, 112*, 605—613.

Marañón, G. (1994). *Viaje a Las Hurdes: El Manuscrito Inédito de Gregorio Marañon.* Madrid: El País-Aguilar.

Mariano-Juárez, L., Rivero-Jiménez, B., López-Lago, L., & Conde-Caballero, D. (2020). Del hambre a la excelencia. Aproximaciones etnográficas a Las Hurdes desde la Antropología de la alimentación. *The Foodie Studies Magazine, 5*, 1—24.

Matías, D. (2017). *La Producción Geosimbólica de Las Hurdes. Teoría, Historia y Práctica de Un Territorio Imaginario.* Universidad de Extremadura.

Matías, D. (2020). *La Leyenda de Las Hurdes. Geografía,* Literatura e Historia de Una Comarca Mítica. Badajoz: Diputación de Badajoz.

Mauleón, J. (2010). Sistema Alimentari Sostenible i Agricultura Familiar: El Pasturatge Al País Basc. In F. X. Medina (Ed.), *Reflexions Sobre Les Alimentacions Contemporánies. De Las Biotecnologies Als Productes Ecológics* (pp. 73—110). Barcelona: Editorial UOC.

Medina, F. X. (2017). Reflexiones Sobre El Patrimonio y La Alimentación Desde Las Perspectivas Cultural y Turística. *Anales de Antropología, 51*, 106–113.

Neumann, C. B. B., & Neumann, I. B. (2018). *Interview techniques. Power, culture and situated research methodology* (pp. 63–77). Springer.

Ottrey, E., Jong, J., & Porter, J. (2018). Ethnography in nutrition and dietetics research: A systematic review. *Journal of the Academy of Nutrition and Dietetics, 118*(10), 1903–1942.

Patrimonio Gastronómico Protegido. (2020). Patrimonio Gastronómico Protegido. 2020.

Rappaport, R. A. (1987). *Cerdos para los antepasados. El ritual en la ecología de un pueblo en Nueva Guinea*. Madrid: Siglo XXI.

Rivero Jiménez, B., Caballero, D. C., & Mariano Juárez, L. (2020). Educación Para La Salud y Alimentación En Personas Mayores. Tradición, Cultura y Autoridad En Las Hurdes (Extremadura-España). *Agathos, Atención Sociosanitaria y Bienestar, 3*, 18–25.

Rivero Jiménez, B., Ortiz, L. L.-L., Conde-Caballero, D., & Mariano Juárez, L. (2020). Del Hambre a La Excelencia. Aproximaciones Etnográficas a Las Hurdes Desde La Antropología de La Alimentación. *The Foodie Studies* (5).

Rivero-Jiménez, B., Conde-Caballero, D., & Mariano-Juárez, L. (2020). Health and nutritional beliefs and practices among rural elderly population: An ethnographic study in Western Spain. *International Journal of Environmental Research and Public Health, 17*(16). Available from https://doi.org/10.3390/ijerph17165923.

Romero, I., Gómez, I. P., & Zabala-Iturriagagoitia, J. M. (2019). 'Cookpetition': Do restaurants coopete to innovate? *Tourism Economics, 25*(6), 904–922.

Sánchez Martín, J. M., & Rengifo Gallego, J. I. (2019). Evolución Del Sector Turístico En La Extremadura Del Siglo XXI: Auge, Crisis y Recuperación. *Lurralde: Investigación y Espacio* (42), 19–50.

Santana, A. (2003). Mirando Culturas: La Antropología Del Turismo. In Á. R. Gil (Ed.), *Sociología Del Turismo* (pp. 103–126). Madrid: Ariel.

Santich, B. (2004). The study of gastronomy and its relevance to hospitality education and training. *International Journal of Hospitality Management, 23*(1), 15–24.

Schmigalle, G. (1993). *Viaje a Las Hurdes: El Manuscrito Inédito de Gregorio Marañón y Las Fotografías de La Visita de Alfonso XIII*. Madird: El País Aguilar.

Serra, J., Correia, A., & Rodrigues, P. M. M. (2015). Tourist spending dynamics in the Algarve: A cross-sectional analysis. *Tourism Economics, 21*(3), 475–500.

Shah, A. (2017). Ethnography? Participant observation, a potentially revolutionary praxis. *HAU: Journal of Ethnographic Theory, 7*(1), 45–59.

Solís, F. (2008). *Buñuel En El Laberinto de Las Tortugas*. Mérida: Editora Regional de Extremadura.

Swain, J. M., & Spire, Z. D. (2020). The role of informal conversations in generating data, and the ethical and methodological issues. *Forum Qualitative Sozialforschung/Forum: Qualitative Social Research, 21*.

Unamuno, M. D. (1922). *Andanzas y Visiones Españolas*. Madrid: Renacimiento.

Velasco, H., & Rada, Á. D. D. (2006). *La Lógica de La Investigación Etnográfica*. Madrid: Trotta.

CHAPTER FOURTEEN

The Organic Market of Mazatlan (Sinaloa, Mexico). Paradoxes of food supply, tourism, and migration

José Antonio Vázquez-Medina, Erika Cruz-Coria, Elizabeth Olmos-Martínez and Mónica Velarde-Valdez
Universidad Autónoma de Occidente, Regional Unity of Mazatlan, Mazatlan, Mexico

Introduction

Historically the State of Sinaloa has been one of the most important suppliers of farm products for human consumption at the national level in Mexico. According to the latest state government report published in 2020, the state remains among the three entities with the highest rates of agricultural production that annually exceed more than 12 million tons of products for consumption nationally and internationally, highlighting the leadership in the production of vegetables, fruits, and vegetables such as cucumber, mango, corn, tomato, and oilseeds such as sesame (Gobierno del Estado de Sinaloa, 2020).

The above figures not only reflect the current agricultural production capacity that concentrates a large part of the national production of certain foods, but also account for the deep-rooted tradition of an intensive agricultural production model in the state since the mid–20th century that has allowed it to be one of the largest exporters of grains and vegetables at the national level, thanks to the successful implementation of the Green Revolution that catapulted it as one of the main suppliers of agricultural products. However, despite the fact that the intensive production model with the exhaustive use of agrochemicals guarantees the production capacity of commercial agriculture in the entity, the agroenvironmental consequences of this model are becoming more and more evident with the problems of contamination and scarcity of water, as well as the

Food, Gastronomy, Sustainability, and Social and Cultural Development.
DOI: https://doi.org/10.1016/B978-0-323-95993-3.00011-6

© 2023 Elsevier Inc.
All rights reserved.

239

persistent trend toward monoculture of grains such as corn, beans, sorghum, and chickpeas (Palau & Trujillo, 2021).

Given the evident deterioration of the Sinaloa countryside due to the conditions mentioned above, within the roadmap of the 2016—20 state administration, several programs were promoted to maintain competitiveness in agricultural production from a sustainable viewpoint. Monetary support for seasonal producers, courses and training for producers on organic agriculture, and support with infrastructure to promote crop diversification were some of the projects that were carried out to mediate the intensive production of industrial agriculture (Gobierno del Estado de Sinaloa, 2020). It is noteworthy, however, that none of the strategies and programs focused on the regional marketing of products to shorten distribution channels, nor were mechanisms put in place to structure the value chain of foods that are produced under such conditions, despite the proximity between the production centers and the final points of sale. At the national and international level, it has been shown that the consumption of locally produced food can be one of the most effective strategies to reduce the ecological damage of their production (Food and Agriculture Organization for the United Nations (FAO), 2016); and the shortening of distribution channels has been prospected as one of the most promising actions for the achievement of sustainable development goals of the 2030 agenda that are related to food.

The lack of revitalization of food distribution chains is more contrasting in the south of the state, where an important part of the state's agricultural production is concentrated and Mazatlán, one of the most important sun and beach tourist destinations in Mexico, is home to. This locality is one of the most densely populated urban centers of the state. Although it is true that in this destination the demand of the national tourist does not necessarily look for the origin of the agricultural products within their tourist experience as an added value to it, but rather the freshness of the sea products and the cultural-traditional load conferred (Flores-Gamboa & Solorzano, 2013), one can also appreciate the emergence of a new profile of consumers—generally permanent residents, temporary residents, and foreign tourists—who are interested in the origin of their food, and that some of them have organized themselves to develop collective initiatives through new spaces where agrochemical-free products are offered in the port. To this contingent of consumers should also be added those producers who offer other types of food such as dairy products, sausages, and artisanal products that use inputs produced under an environmentally friendly scheme and that have sought the opening of spaces where they can market their products.

This new wave of consumers can be explained by the population recomposition of Mazatlán registered for several decades, which has to do with the residential tourism of foreigners permanently settled or itinerant in the port (Lizárraga, 2016; Maldonado, 2017).

Despite the fact that Mazatlán has become a tourism alternative for American vacationers since the first half of the 20th century (Santamaría, 2009), it was not until a couple of decades ago that the port began to position itself as one of the favorite destinations as second homes for retirees from the United States and Canada to overcome the winter months (Lizárraga, 2016; Maldonado, 2017). Being a destination that offers a benevolent and affordable climate for international tourists, little by little Mazatlan began to become a circular migration pole that houses, mostly, older adults from the countries to settle for extended periods of time; weather or, to establish their permanent or semipermanent residence in the port.

According to Lizárraga (2008, 2016), Ceballos (2021), Maldonado (2017), and Sandoval-Pintor (2020), the arrival and settlement of American migrants in Mazatlán have modified the spatial and consumption dynamics in some areas of the city, expanding new alternatives aimed at this sector, but also unbalancing the regulation of the housing supply market, products, and services. These readjustments in the population dynamics from the arrival of the new inhabitants can also be made visible through the transformation of the local foodscape with the opening of new consumption spaces that are aimed at this segment of the population.

Other texts have discussed how mobility phenomena such as migration and tourism structurally condition urban foodscapes based on a recalibration of food supply (Vázquez-Medina, 2018, 2021). In the case of Mazatlán, this reconfiguration is present not only in the points of sale of ready-to-eat food, but also in the supply points that offer specialized products for new consumers, made up mostly of Americans and Canadians who are permanent migrants, temporary migrants, and/or tourists who have settled in the city. One of the supply centers that confirms the above is the Mazatlán Organic Market (MOM): a mobile market that offers organic products installed temporarily during the winter months in the central area of the city.

The MOM is located in Plazuela Zaragoza, a strategic point in the historic center, which also represents an important part of the neighborhood transformation that has been given to this area for tourist purposes. The market has become a recurring space for foreign retirees who live in the port during this season, and for some locals who are looking for other, healthier food options. It has also been included in the recommended itineraries for this sector of visitors as a means of tourist promotion.

Although these types of initiatives make food production and supply chains more dynamic, shortening the distance between producers and consumers, it is also true that the segmentation of the market to which they are directed restricts access to these foods to other sectors of the port's inhabitants due to the sale prices and the formats in which the products offered there are sold. In this way, the MOM materializes some important paradoxes regarding access by the local population to fresh products free of agrochemicals, which have been discussed within studies of the sustainability of contemporary food systems: on the one hand, this type of product enables a new marketing space by dynamizing the supply chain. On the other hand, given the difference in consumer prices for organic versus nonorganic products, the target market is highly segmented toward foreign migrants and tourists with ample purchasing power who demand this type of product during their stay in the port, restricting access to other sectors of the Mazatlecan population.

Due to the above, the objective of this chapter is to analyze the relationship between migration and tourism as mobility phenomena that restructure the context of food production and distribution in the city of Mazatlán and reflect on the impact of this dynamic on the promotion of more sustainable and resilient food systems.

The empirical findings of this research were recovered through a combination of various qualitative techniques in fieldwork that took the ethnographic method as the basis for their realization. For this, participant observation was carried out in the MOM, between November 2021 and April 2022, to cover the entire eleventh season of the same.

The visits included, for the most part, the observation of the entire operating hours of the market; that is, from eight in the morning until noon. On some occasions, the spectrum of observation was opened to analyze the installation and removal of stalls. The observation was complemented by fifteen semistructured interviews and informal talks with the key actors involved in the market: managers/administrators, producers, sellers, and national and foreign consumers. Consent for the interviews was given verbally prior to conducting the interviews.

Conceptual approaches to the configuration of new consumers in the framework of migration and tourism

The present work uses the anthropology of food to analyze how social processes of a macro order such as tourism and migration are capable of intervening in the phases of the local food system that have to do with provisioning and consumption to make the foodscape more complex. To explain this, some theoretical approaches from human geography, cultural and urban sociology, and tourism studies interested in the sociocultural aspects of food were included.

On food supply and consumption, this text considers discussions that have been scrutinized to characterize the profile of new consumers who are interested in the origin of their food and the way it is produced and that their emergence can be understood as a consequence of exhaustion of contemporary food systems. "New consumers," "reflective consumers," "globalized consumers" or "consumactors," are some of the names that have been created from the field of social and cultural food studies to characterize the profiles of consumers that have emerged in response to *food modernity*, restructuring the relationship between subjects and their food in recent decades (Boucher, 2019; Espeitx, 1996; Fischler, 1995; Gracia, 2005). Almost as a consensus among the authors, a consumer–citizen has been outlined who, according to their purchasing power and cultural capital, tries to face post-Fordist food production through selective consumption that favors a friendly and comprehensive relationship with the consumer environment that provides them with food, while paying attention to community aspects inherent to their production. In this sense, as mentioned by Medina and Aguilar (2022), the emergence of a new consumer arises from the need to provide some certainty about the origin of their food in the face of their growing industrialization processes. In the Mexican case, some incipient studies also relate this type of consumption to aspects related to health care in the face of disease processes (Osorio, 2020; Pardo & Durand, 2019; Vázquez-Mendoza, 2018).

The characterization of the profiles of new consumers has also reached an area of interest for this study that is becoming more and more widespread: sustainable consumption. According to Medina and Aguilar (2022), the opening of new supply spaces for these new consumers has contributed, albeit in a very incipient way, to initiating a transition towards more sustainable food systems. However, from a sociocultural point of view, Torres Salcido (2018) agrees with Johnston and Baumann (2015) in warning that the emergence of these new consumers and the generation of new supply spaces where this type of product is marketed also contributes to promote inequity in access to food with these characteristics and elitist its consumption, triggering other imbalances in urban food systems. In this sense, the reflections of Jover and Díaz-Parra (2020) are also of interest to this study to analyze how the arrival of tourists or foreign residents to receiving communities favor dynamics of access to activities and infrastructures that exclude the inhabitants of the receiving communities.

In relation to the interconnection processes between migration and tourism, the epistemological discussions about their imbrications proposed by Williams and Hall (2000) are useful to understand how these phenomena intersect and reinforce each other. In particular, their reflections serve to analyze how the enclave of residential tourists in Mazatlan has been formed to elucidate how tourism can be a mechanism of attraction that favors certain types of temporary or permanent migration, and how this phenomenon perpetuates mobility flows between people. For the case study of this work, the conceptual clarifications proposed by Peraza and Santamaría (2017) on the ethnic economies of Americans in Mazatlán were also recovered to argue that in the case of the population recomposition of the port, tourism was a decisive factor in establishing the ethnic enclave of American retirees and how this dynamited the start-up of some businesses aimed at members of said enclave.

On the other hand, reflections on the sustainability between these two phenomena are also considered, proposed by Choe and Lugosi (2021) who suggest the need to establish critical views related to the asymmetric relationships that are generated in the receiving society when these two phenomena converge in a given territory, and how the interaction between both affects the population recomposition favoring the quality of life of only certain sectors.

Finally, it is agreed that migration and tourism also involve the generation of spaces aimed at migrant groups that can be considered as transnational social spaces (Hirai, 2009), which serve to recreate and reproduce food consumption practices in the communities of reception, at the same time serve as references to explain the complex relationship that is forged between tourism, migration and food (Vázquez-Medina, 2016).

New markets, new consumers, and the generation of emerging food supply spaces in Mazatlan

For eleven years the MOM has been held regularly on Saturday mornings, between November and April, in the season of greatest reception of migrants from northern countries called winter birds or snowbirds, which also coincides with the harvest of most of the products offered on the site. Originally, the market was an initiative of an American journalist who, upon retiring, changed her residence to Mazatlán and was looking for some fresh products that were free of agrochemicals to maintain the consumption patterns and habits that she had before migrating to the port. According to the testimony of the current administrator:

> The founder is originally from the United States, from California. And in California the farmers' markets are very common. For the same reason that people seek more freshness, more natural, less chemicals; So, the idea and concept of the market comes from there, of being able to find the same thing that they consumed before in their new place where they came to live. (Personal communication. February 2022)

From the studies of migration and food, the idea has been generated that in displacements there is a generalized desire to maintain a food *continuum* in the communities of reception, which also implies the perpetuation of the dynamics of consumption in the societies of arrival (Parasecoli, 2014). With this purpose in mind, the newcomers are capable of creating and mobilizing structures to be able to maintain and perpetuate the eating habits and patterns that they had before migrating. As stated by Clifford (1999), in the phenomenon of human mobility, traveling cultures favor the diversification of certain consumption practices in the societies of their arrival.

The case of the creation of the MOM is especially illustrative in corroborating these affirmations. Once rooted in the port, the founder of the market contacted chefs and business owners who were interested in offering this type of product and jointly launched a call that was disseminated through social networks aimed at producers who grow food under a chemical-free scheme. According to the current market managers, five or six producers met in the first edition of the market. Currently, the MOM is an association made up of more than 30 members who bring together producers of vegetables, fruits, dairy products, and eggs, as well as sellers of personal care products, artisanal bakeries, and other handicrafts.

Regarding the way of operating and managing, each member pays an annual membership that allows them to settle during the entire season that the market lasts, and although there is no current verification mechanism by the association that guarantees organic production, there is a regulation that stipulates the conditions of participation. One of the aspects included in the operating rules is the balance in the supply of the types of products that are sold in the market. In other words, efforts are made to ensure that there is a similar number of points of sale between fresh agricultural products, artisan products such as sausages, dairy products, and preserves, and other types of products such as handicrafts and decorations that are currently sold in the market. As reported by the current administrators, in recent years, it has been difficult to find producers who sell organic fruits and vegetables who decide to participate, which has limited their possible expansion into other types of products.

According to the data provided by the current administration, each year the MOM receives a larger influx of clients. In its last edition, the MOM hosted every Saturday, and had an average of between 300 and 400 visitors. Of the total number of visitors more than 70% are foreigners. With the intention of encouraging the attraction of clients, every Saturday at the MOM there are also free activities such as yoga and dance classes, as well as culinary demonstrations, gardening workshops, and other recreational activities. The MOM has tables in its facilities for the congregation of its visitors, who, for the most part, have been loyal customers not only for the products that are offered but also for the social and recreational space that it represents.

During interactions between customers and suppliers, it is common to hear aspects of food production mentioned as a decision criterion to guarantee the purchase. Isabel, a producer who sells vegetables and preserves in the market, says that more and more Americans try to know the conditions under which their food is produced. Among the concerns most commonly referred to are the mode of food production that were recovered in the fieldwork, the type of fertilizers used in production, the safety of the water used to irrigate the inputs, and the conservation of crops stand out, as do the products during the transfer from the communities where they are produced to the port of Mazatlan.

Regarding the selection criteria of the products, the interest generated in the way of production and the commodification of the proximity of the same that are made visible in the promotion of the market stalls, contribute to the emergence of a new profile of consumers interested in knowing the conditions in which their food is produced. Faced with the hyper-homogenization of global food systems, new consumers appear who have the privilege of attributing certain moral values to food. These values, as Contreras and Gracia affirm, fall on solidarity, green consumption, the discourse of sustainability, fair trade, and balanced and healthy products (2005:443). Against this background, what is fresh and local and within a localized territory, have become attributes that are increasingly sought after among contemporary consumers.

These new consumers have been able to translate production knowledge into gastronomic values that guide their food decisions and generate added value for the products that are marketed; and as Pardo and Durand (2019) affirm, the commodification of nature expressed in this type of space visibly resignifies the relationship with food and confronts the global standardized diet, although its scopes are clearly sectorized.

Between the offer, the restriction, and the added value of organic products in the Mazatlán Organic Market

Cecilia is a neighbor who has lived for more than twenty years in the vicinity of Plazuela Zaragoza. She attends the market as a walk to distract herself on Saturday mornings, or to have breakfast while listening to live music she likes. In her words, "it is impossible to buy things for the week here. If anything, some other Saturday (I buy) bread, or honey, which is of the best quality. Here it seems that they translate the price of food into dollars" Ricardo, a university student, travels almost an hour every Saturday during the MOM season to buy his fresh food at the market, which implies a significant readjustment in his weekly budget and the organization of his activities to acquire food. For Ricardo, the consumption of this type of food is beneficial for health, and he also agrees that the producer receives a fair price for what he sells, even if that means paying more. Although he also adds that this type of offer restricts access to other sectors of the population:

> *It is good that there are these options, even if it is during the winters. It is also good that foreigners come to spend their dollars here, because that implies that they go directly to the producers, that there are no intermediaries. Yes, it is more expensive, but you start to think that the benefit is for the producers and*

that is already worth the investment, although it is rather a kind of luxury...
(Personal communication. March 2022)

The above perceptions are contrasted with the view of Lindsay, one of the market administrators, who states that the selling price of products is becoming more accessible to the general public:

Before, the organic, the natural, and the expensive were closely related. Now that is not true. If I go to Walmart and stock my fruit and vegetables there, you will see that the difference is not that great and the quality is never compared to what you can find here (Personal communication. March 2022).

Lindsay's testimony agrees with the assessments of Anna and Sam, a retired couple from Minnesota who have settled for three months in Mazatlan for the third year. For them, the prices are extremely competitive in relation to the quality and freshness of the products, compared with what they can find in a supermarket and, above all, compared with the prices at which they usually buy in their usual place at home:

We like to come here for the quality and freshness of the food you find. The lettuces are fantastic, much better than in the supermarket. Where we come from, it would be impossible to find such fresh, organic lettuces and leafy vegetables at this price. We also take the opportunity to spend some time and meet with friends who come here to the market (personal communication, March 2022).

The difference in the perceptions retrieved here accounts for the evident restriction in access to the products offered in the MOM. Although these types of exercises involve collective action to bring producers and consumers closer and, finally, benefit the former with a fair price, the truth is that this is not always the case. Víctor is a producer who, after working for a fertilizer company for several years and knowing in depth the damage caused to products by the abuse of these substances in conventional agriculture, decided to start growing organic products in a town about 80 km from Mazatlan. He is aware that the sale price is higher than its counterpart that offers commercial agricultural products and that it represents a significant expense for any family budget of the average local population, but he refers to the fact that, even for those who try to distance themselves from the commercial logic of commercial agriculture, it is impossible not to depend on the latter:

We also suffer from changes in prices, because apart from that who is dedicated to agriculture knows how things are. A few weeks ago, in January, in the field they were paying one peso per kilo of tomatoes, chili and tomatillo tomatoes. That automatically lowers the price in any market and makes what we bring more expensive for people. And it happens that we have a lot of products left that we can no longer move and sell because people find the price difference greater. But then those who do sell their product in the field at that price begin to freak out, because it is impossible to sell at that price and be able to maintain it, no matter how you produce; and then the price begins to rise in other places and is a little closer to what we have and maintain throughout the season, but we will always be conditioned to what happens there, but we know that the prices we offer do affect the household economy... (Personal communication, February 2022).

The previous testimonies allow us to understand that the restriction on access to this type of product can be understood as a multicausal relationship and not as a conscious action to exclude other sectors of the population. On the part of foreign consumers, their profile, which supposes a high purchasing power that benefits from the Mexican peso-dollar parity, allows the loyalty of their supply regardless of the sale price of the products offered in the market. For national consumers, being able to purchase any of the products implies a significant readjustment in the budget for food spending, which causes another more restrictive segmentation of access to this type of product by the local population. On the producers' side, the difficulty of competing with the prices imposed by the logic of the conventional agriculture market increases segmentation in the types of profiles that can access these products.

In this sense Pardo and Durand (2019) agree with Joassart-Marcell and Bosco (2017) that these types of markets can be assumed as collective actions that challenge the capitalist logic of food distribution and consumption that regulates contemporary urban food systems. However, in the case of the MOM, due to other broader social processes such as migration and tourism, the logic of the market reproduces many of the dynamics that it wants to neutralize and are part of the neoliberal system of food production and consumption.

From culinary cosmopolitanism to agriculture on demand: new products and culinary information flows in the Mazatlán Organic Market

Isabel and her sister grow crops in the Barras de Piaxtla área, north of Mazatlán. Among the products sold in the MOM this season, the Pak-Choi, a variety of cabbage that is widely used for the preparation of oriental stews, stands out. As reported by the sisters, an American client brought them some seeds two years ago so that they could plant them, since it was a food widely consumed by them in the United States and was not found in Mazatlan. Víctor, for his part, produces and sells a variety of plantain called *Plátano Thai* that is not marketed in Mexico, but rather, due to its high demand, is exported directly to the California region in the United States. At the MOM, he has found a market niche that is in great demand for this type of banana. Jorge, who owns a tomato greenhouse near San Ignacio, produces tomato varieties such as the Green Zebra, German Jhonson, and Kentucky that are not of common culinary use in Mexico, but that their foreign clients demand at the MOM.

Previous experiences corroborate the notion of markets as transnational spaces that generate flows and information about merchandise for culinary use. In this sense, it is undeniable that the MOM can be considered an empowerment center for culinary cosmopolitanism that is generated thanks to the clientele made up of migrants who take root in Mazatlan temporarily and tourists who visit the port. For this reason, the MOM represents a center for the irradiation of culinary information that can diversify food consumption with products that can be easily cultivated, counteracting, albeit on a small scale, the tendency toward a monoculture mentioned at the beginning of this chapter.

However, it can be seen that the relationship between the producer and the consumer has generated a degree of dependency that conditions the type of product to be offered, which is aimed only at a certain sector of the population, generating production on demand, contributing to hyper-segmentation of the market, and restricting the widest access to these types of products. As Jorge, an agricultural engineer, who grows different types of chili peppers, stated:

> We are trying to reduce the spiciness of the jalapeño pepper for the next crop season because most of our American clients complain about it. We are exploring how to do it and I think we can make it. They love the flavor but, for them, the spiciness is too much. Also, we will try to grow other varieties of chili that are less spicy (Personal communication, March 2022).

It would then be worth questioning whether these relations between producers and consumers, which are free of intermediaries, but generate a certain level of dependency and condition the type of crops, really contribute to overcoming the depletion of intensive production that afflicts

The Organic Market of Mazatlan (Sinaloa, Mexico)   251

the local food system, and to what extent they contribute or not to its sustainability.

Final considerations

This chapter has studied some of the constraints that link the production and consumption of food with broader social processes that have to do with mobility, such as migration and tourism. Due to its characteristics, the MOM challenges some preconceptions about the type of exercises that alternative markets engage in as collective actions contributing to the generation of sustainable strategies that address the problems of contemporary food systems. Although it is true that this market efficiently stimulates supply chains with differentiated quality products that can help the food transition to a sustainable food system, it is also evident that due to the very nature of its composition, it triggers other order problems, such as social, economic, and cultural.

In this sense, it is important to note that even self-managed citizen initiatives that challenge conventional models of food production and distribution, contribute to favoring or accentuating other problems that impact urban food systems such as food justice, the elitization of locally produced food, and food gentrification. The analysis of the MOM also shows the need to include within the strategies to counteract the exhaustion of contemporary food systems, the interaction that this has with other broader social processes that affect the contexts of the territories where they are carried out, such as migration and tourism, to make these types of exercises contribute to the food transition toward fairer, more sustainable, and more resilient food systems.

References

Boucher, F. (2019). Construcción del enfoque sial, la Red sial México y los nuevos conceptos. In M. C. Renard, & J. M. Tolentino (Eds.), *Red sial México Diez años de contribución a los estudios de los Sistemas Agroalimentarios Localizados* (pp. 13−33). Mexico City: CONACYt-Red sial-México-ICCA.

Ceballos, T. (2021). Los desafíos del turismo en Mazatlán. Hacia una nueva estrategia de desarrollo local. In E. Hernández, & G. Ibarra (Eds.), *Los Grandes Problemas de Sinaloa* (pp. 45−76). Mexico City: Tirant Lo Blanch.

Choe, J., & Lugosi, P. (2021). Migration, tourism and social sustainability. *Tourism Geographies*, *24*(1), 1−8. Available from https://doi.org/10.1080/14616688.2021.1965203.

Clifford, J. (1999). *Itinerarios transculturales*. Barcelona: Gedisa.

Contreras, J., & Gracia, M. (2005). *Alimentación y cultura. Perspectivas antropológicas*. Barcelona: Ariel.

Espeitx, E. (1996). Los nuevos consumidores o las nuevas relaciones entre campo y ciudad a través de los «productos de la tierra». *Agricultura y Sociedad*, 83—116, 80—81.

Fischler, C. (1995). *El (h)omnívoro. El gusto, la cocina y el cuerpo*. Barcelona: Anagrama.

Flores-Gamboa, S., & Solorzano, C. A. (2013). Perfil del turista nacional que consume alimentos durante Semana Santa en Mazatlán, Sinaloa. *Teoría y Praxis, 13*, 59—81.

Food and Agriculture Organization for the United Nations (FAO). (2016). *Memoria del taller de intercambio de experiencias en Cadenas Cortas Agroalimentarias*. Ciudad de México.

Gobierno del Estado de Sinaloa. (2020). *Cuarto Informe de Gobierno de Quirino Ordaz*. Retrieved from http://estadisticas.sinaloa.gob.mx.

Gracia, M. (2005). Maneras de comer hoy. Comprender la modernidad alimentaria desde y más allá de las normas. *Revista Internacional de Sociología, 40*, 159—182.

Hirai, S. (2009). *Economía política de la nostalgia. Un estudio sobre la transformación del paisaje urbano en la migración transnacional entre México y Estados Unidos*. México Ciudad de México: UAM Instituto Mora.

Joassart-Marcell, P., & Bosco, F. (2017). *Food and Place. A Critical Exploration*. Lanham: Rowmand and Littlefield.

Johnston, J., & Baumann, S. (2015). *Foodies. Democracy and Distinction in the Gourmet Foodscape*. Nueva York: Routledge.

Jover, J., & Díaz-Parra, I. (2020). Who is the city for? Overtourism, lifestyle migration and social sustainability. *Tourism Geographies, 24*(1), 9—32. Available from https://doi.org/10.1080/14616688.2020.1713878.

Lizárraga, O. (2008). La inmigración de jubilados estadounidenses en México y sus prácticas transnacionales, Estudio de caso en Mazatlán, Sinaloa y Cabo San Lucas, Baja California Sur *Migración y Desarrollo* (11), 97—117.

Lizárraga, O. (2016). El Turismo Residencial de Retiro en Mazatlán, México. Una propuesta de marca ciudad Revista Anais Brasileiros de. *Estudios. Turístico/ABET, Juiz de Fora, 6*(3), 85—96.

Maldonado, C. M. (2017). La Competitividad Turistica de los Destinos Costeros en Mexico ante el Mercado de Turismo Residencial: el caso de los norteamericanos en Mazatlán/Sinaloa. *Anais Brasileiros De Estudos Turísticos, 7*(3), 74—87. Available from https://doi.org/10.34019/2238-2925.2017.v7.3187.

Medina, F. X., & Aguilar, A. (2022). ¿Hacia sistemas alimentarios sostenibles en contextos culturalmente coherentes?; debates en torno a la cultura, la sostenibilidad y la dieta mediterránea. In D. Conde, L. Mariano, & F. X. Medina (Eds.), *Gastronomía, cultura y sostenibilidad. Etnografías contemporáneas* (pp. 25—40). Barcelona: Icaria.

Osorio, L.C. (2020). *Cooperativas chinamperas y canales cortos de distribución de alimentos en la Ciudad de México*. BA Dissertation in Gastronomy Studies. Universidad del Claustro de Sor Juana.

Palau, E., & Trujillo, J. D. (2021). Los problemas centrales de la agricultura comercial en Sinaloa y la soberanía alimentaria. In E. Hernández, & G. Ibarra (Eds.), *Los Grandes Problemas de Sinaloa* (pp. 15—44). Mexico City: Tirant Lo Blanch.

Parasecoli, F. (2014). Food, Identity, and Cultural Reproduction in Immigrant Communities. *Social Research, 81*(2), 415—439.

Pardo, N., & Durand, L. (2019). Consumir y resistir: los mercados alternativos de alimentos en la Ciudad de México. In L. Durand, A. Nygren, & A. De la Vega (Eds.), *Naturaleza y Neoliberalismo en Iberoamérica* (pp. 467—504). Mexico City: Uiversidad Nacional Autónoma de México.

Peraza, B. E., & Santamaría, A. (2017). ¿Turistas o inmigrantes estadounidenses? Identidad y economías étnicas en Mazatlán, Sinaloa (México). *Revista Latino-Americana de Turismologia*, *3*(2), 24−37.

Sandoval-Pintor, R. (2020). Breve radiografía de la (in) migración interna y segmentación laboral de Sinaloa. *Ra Xhimai*, *16-1*, 125−152.

Santamaría, A. (2009). *El Nacimiento del turismo en Mazatlán*. Culiacán: Editorial Universidad Autónoma de Sinaloa.

Torres Salcido, G. (2018). Gestión y gobernanza territorial. Los Sistemas Agroalimentarios en la encrucijada del desarrollo territorial. *Revista Iberoamericana de Viticultura Agroindustria y Ruralidad*, *5*(14), 65−79.

Vázquez-Medina, J. A. (2016). *Cocina, nostalgia y etnicidad en restaurantes mexicanos de Estados Unidos*. Barcelona: Editorial UOC.

Vázquez-Medina, J. A. (2018). La emergencia de turismo gastronómico «étnico» en comunidades rurales a partir del retorno de migrantes: *Comer chino* en Santa María del Rio, S.L.P. In M. P. Leal, & F. X. Medina (Eds.), *Gastronomía y Turismo en Iberoamérica* (pp. 159−171). Gijón: Editorial Trea.

Vázquez-Medina, J. A. (2021). Migración de retorno, alimentación y territorio en municipios semirrurales del altiplano norcentral de México. In Y. Peña, & L. Hernández (Eds.), *Tradición y Patrimonio. De la historia a los escenarios globales* (pp. 205−222). Mexico: Instituto Nacional de Antropología e Historia.

Vázquez-Mendoza, D. (2018). *Etiqueta Chinampera: una herramienta para la conservación de Xochimilco y el axolote*. Mexico City: Facultad de Ciencias, Universidad Nacional Autónoma de México, Mexico City, Mexico. BSc Dissertation.

Williams, A. M., & Hall, C. M. (2000). Tourism and migration: New relationships between production and consumption. *Tourism Geographies*, *2-1*, 5−27.

Index

Note: Page numbers followed by "*f*" and "*t*" refer to figures and tables, respectively.

A

Acculturation, 220
Activities, 110−111, 193
  agricultural, 217−219
  food-related, 159−160
  gastronomic, 10−11
  recreational, 178−179
  smallholder, 17−18
  tourism, 187
AFNs. *See* Alternative food networks
  (AFNs)
Agency, 86−87, 120−121, 127
Agenda 2030
  gastronomic tourism and, 189−190
    alternative food networks, 186−189
  Pyrenees tourism brand, 190−194
Agri-food
  products, 10
    of region, 179−180
  routes, 170, 172, 178
    and sustainable rural development,
      170−172
  sustainability of agri-food systems, 171
Agricultural activities, 217−219
Agricultural production
  chains, 1
  challenge, 220−221
  traditional, 234
Agricultural products, 36, 85
Agricultural sector, 178−181
Agriculture, 18−19, 32, 187−188,
    220−221
  conventional, 217−219
  from culinary cosmopolitanism to
    agriculture on demand, 249−251
Agro-nutritional system, 49−50
Agrochemicals, 242
Agroecosystem, 178−179
AI. *See* Artificial intelligence (AI)
Alternative food networks (AFNs), 5,
    10−11, 185−186, 188

Anthropomorphism, 100
AO. *See* Appellations of origin (AO)
Appellations of origin (AO), 193
Arroz a la cazuela, 125
Artificial intelligence (AI), 8, 95, 104
Artisan liqueurs, 208
Artisan(al) products, 229−230, 240−241,
    246
Artistic critique, 134−135
Aspirational consumption, 7−8, 87−91
Association of Sustainable Restaurants,
    122−123
Asylum-seekers reception centers in Berlin,
    157−158, 159*f*
Automatic service delivery, 107−108
Autonomous Community of Catalonia,
    121−122
Autonomous Community of Extremadura,
    201, 207

B

Barcelona, 121−123, 128−130
Beans, 239−240
Beef consumption, 86
Berlin, asylum-seekers reception centers in,
    159*f*
Biodiversity, 34−35, 137−138, 148−149
Black legend of Las Hurdes, 201−204
Brazil, meat consumption in, 86−89
British allotment systems, 16−17
Buñuel's images, 202−203
Buñuelos, 208

C

Capital accumulation process, 134−135
Capitalism model, 134
Capsicum annuum. *See* Yahualica chilli
    (*Capsicum annuum*)
Capsicum chinense. *See* Habanero (*Capsicum
    chinense*)
Carbon emissions, 26

Carbon footprint, 4–5, 77–78
Catalan cuisine, 123–126, 128–130
Catalan culture, 129–130
Catalan products, 130
Catalan recipes, 125
Catalan-style dish, 130
Catering, 151–153
Chef, 54, 120–121, 143, 245
Cherries, 208
Chestnut panna cotta, 208
Chestnuts, 208
Chickpeas, 239–240
Chile Yahualica, denomination of, 172–175
Chili peppers, 250
Chili production, 173
China, meat consumption in, 86–89
Civic engagement through food, 151–154
Civil society, 122, 155–156
Climate change, 1, 26, 32, 49
Codex Alimentarius, 227–228
Communicative system, 151–152
Community gardens, 16–17
Constant process of reinterpretation, 124
Construction process, 120
Consumers, 240–241
  experiences, 99–100
  of organic products, 232
Consumption chain, 117
Consumption power, 85
Contemporary debate, 5–6
Convention theory, 136t
Conventional agriculture, 217–219
Conventional food systems, 10–11
Cooking, 9–10, 151–152
  classes, 163–164
  community, 163
  techniques, 77, 129
Cooks, 146
Corn, 239–240
Cosmopolitan city, Barcelona as, 128–130
Covid-19, 26–27, 95
  pandemic, 96, 101, 103, 154, 205
  tourism, 95
CRESIAP. *See* Regional Center for Integral Services for Protected Agriculture (CRESIAP)

Crop rotation system, 174
Cucumber, 239
Cuisine collectives, 193
Culinary cosmopolitanism, 249–251
Culinary system, 2, 37–40
Culinary tourism, 209–210
Cultivation, 22–23
  plots, 173–174
Cultural diversity, 163
Cultural identity, 123–126
Cultural materialism, 215–216
Cultural production, 137–138
Cultural resources, 1
Cultural values, 187–188
Cultural-gastronomic-commercial current, 144
Culture, 2, 90–91, 187–188
Cured meats, 208
Customers without intermediaries, 146

D
Dairy products, 240–241, 246
Danish restaurant, 133
Decorations, 246
Deforestation, 84
Delicatessen, 231–232
Delivery processes, 21–22
Denomination of Chile Yahualica, 172–175
Denomination of Origin (DO), 169
Deprivation, abundance *vs.* narrative of, 206–208
Devouring nature without feeding people, 217–221
Diets, 32. *See also* Sustainable diets
DO. *See* Denomination of Origin (DO)
Dragor coast, 140–141

E
EAT-Lancet Commission, 84
Economic values, 187–188
Economies, 16–17, 20–21
Education, 151–152
EEC. *See* European Economic Community (EEC)
Electric engines, 25–26
Electric motor, 155–156

Electricity, 25–26
Emission of greenhouse gases, 85
Energy, 5–6, 15–16
  consumption, 5–6, 35
Environmental impact, 83, 88, 91
  of meat, 84–85
Environmental risk, 33
Environmental sustainability, 3–4, 33,
  91–92
Escudella, 125
Ethnic food lifestyles, 52
Ethnic identity, 123
Ethnographic analysis of dietary ideologies,
  205
Ethnographic material, 205
Ethnographic method, 242
European Economic Community (EEC),
  222–223
European Union, 222–223
Exquisiteness, 208–211

F

Fair trade, 10–11, 188, 229–230
Fairs, 229
Falafel, 154
FAO. *See* Food and Agriculture
  Organization (FAO)
Farm-to-fork ingredients, 209
Farmers, 54–56
  markets, 23–24, 188, 193, 245
Farms, 17–18, 23–24
Fast–food restaurants, 99, 104–106
Fats, 39
Feeding concept, 223
Fertilizers, 3
Floretas, 208
Food and Agriculture Organization (FAO),
  2–3, 240
Food bikes, 154
  to bypass immigration structures,
    154–165
  shaping food affects through mobility,
    159–165
    Kitchen on the Run 2016 Europe
      tour, 163f
Food consumption, 5–6, 206
  in city of Lugo, 221–228

and sustainability in Lugo
  contracts, 229–233
  devouring nature without feeding
    people, 217–221
  from global to local, 221–228
Foods, 1, 86–87, 118, 151–152,
  223–224, 240–241, 245
automation in food experiences,
  98–99
chains, 2, 83
continuum, 245
culture, 51–52
distribution chains, 240–241
gastronomic tourism and alternative food
  networks, 186–189
goods, 176
heritage, 50
  movement, 49
industry, 31–32
knowledge, 87
modernity, 243
preparation, 95–96
production, 49–50, 83, 117, 151–152
resources, 215–216, 223
routes, 172
safety, 137–138
scarcity, 89–90
security, 19
service industry, 102
sourcing, 58
stability, 31
supply, 85
  chains, 2, 21–22
  and consumption, 243
  for food service, 52–53
  spaces in Mazatlan, 245–247
  system, 10–11
sustainability, 8–9, 117–118
systems, 31–32
tourism, 98–99
transition process, 119–120
transportation, 16
waste, 4, 31–32, 96–97
on wheels
  forced migration and civic
    engagement through food,
    151–154

Foods (*Continued*)
  pedaling food bikes to bypass
    immigration structures, 154—165
Forced migration, 151—154
Fossil fuels, 25—28
Fresh food, 16—17
Fruits, 18, 239

G

Galician culture, 223—224
Gardens, 5—6
Gastronomes, 4
Gastronomic approach, 209
Gastronomic decolonization, 6—7
Gastronomic excellence, model of, 139
Gastronomic experiences, 96, 100
Gastronomic explorers, 141
Gastronomic festivals, 229
Gastronomic field, 3
Gastronomic industry, 1—2
Gastronomic routes, 193
Gastronomic tourism, 10—11, 185—186
  and Agenda 2030, 189—190
  and alternative food networks, 186—189
  key elements linked to territorial
    sustainability due to, 192t
  research, 187
  supply chain, 186
Gastronomization, 50
Gastronomy, 1, 15, 23—24, 31—32, 50,
  76—77, 177—179, 186, 205—211
Gastroteca, 191
Generic productions, 9
Genetic diversity, 172
Genuine gastronomic diversity, 22—23
German Jhonson (tomato varieties),
  249—250
GHG emissions. *See* Greenhouse gas
  emissions (GHG emissions)
GISCSA. *See* Interdisciplinary Group for
  Studies on Society, Culture, and
  Health (GISCSA)
Global debate, 7—8
Global ecological footprint, 83
Global environment, 16
Global supply networks, 85
Global warming, 20—22

Globalization, 6—7, 49, 220
Gourmet, 231—232
Green agricultural land, 21—22
Green Revolution, The, 217—220,
  239—240
Green Zebra (tomato varieties),
  249—250
Greenhouse fruit and vegetable production,
  41—42
Greenhouse gas emissions (GHG
  emissions), 5—6, 21—22, 35, 84
Greenwashing processes, 5
Group of anthropologists, 204
Growth hormones, 217—219

H

Habanero (*Capsicum chinense*), 169
Halloumi sandwiches, 154
Handicrafts, 246
Haute cuisine, 179—180
Healthy dietary model, 39
Healthy nutrients, 39—40
High-performance genetically modified
  crops, 217—219
High-yielding variety (HYV), 217—219
Homegrown food, 19
Honey, 207—208
Honey orujo, 208
Human library, 163—164
Human-robot interactions, 99—101
  in gastronomic experiences, 100—101
Humanoid service robots, 100
Hurdano gastronomy, 203
Hurdano lemon, 208
Hurdano model, 209—210
Hydropower sources, 25—26
Hypermarkets, 21—22
HYV. *See* High-yielding variety (HYV)

I

ICAF. *See* International Commission on
  Anthropology of Food and
  Nutrition (ICAF)
IMPI. *See* Mexican Institute of Industrial
  Property (IMPI)
Industrialization process, 4, 49
Ingredients, 39—40, 50, 57

Index 259

Intensive agricultural production model, 239–240
Interconnection processes, 244
Interdisciplinary Group for Studies on Society, Culture, and Health (GISCSA), 204
Intergenerational discussion, 89–91
International Beekeeping and Tourism Fair, 207
International Commission on Anthropology of Food and Nutrition (ICAF), 15
International Federation of Robotics (IRF), 97
Internationalization, 220
Interview process, 205, 209
IRF. *See* International Federation of Robotics (IRF)
Irrigation systems, 217–219

J
Jugus curinus, 208
Justification social actors, 136

K
Kentucky (tomato varieties), 249–250
Kitchen container, 161–165
Kitchen scraps, 19

L
Las Hurdes
abundance *vs.* narrative of deprivation, 206–208
construction of "black legend" of, 201–204
exquisiteness, sustainability, and new gastronomy, 208–211
methods, 205
Legendre's thesis, 202–203
Lévi-Strauss formula, 215–216
Lindsay's testimony, 248
Local food, 187–188, 190
Local organic ingredients, 121
Local produce cuisine, 209
Local producers, cooperation with, 126–128
Localism, 15, 23–26

vegetables on Paris market stall, 24f
wild mushrooms from Cerdanya valley, 25f
Lugo, food consumption in, 221–228

M
Macarrones, 125
Malnutrition, 32
Mango, 239
Mann-Whitney test, 106–107, 108t
Marketing resources, 225–226
Markets, 229
Maslow's pyramid of needs, 34
Mass-market development, 87
Matanzas, 206–207
Material artifacts, 153–154
Material dimension, 137–138
Material elements, 118
Mazatlán Organic Market (MOM), 12–13, 241
configuration of new consumers in framework of migration and tourism, 243–244
new markets, new consumers, and generation of emerging food supply spaces in Mazatlan, 245–247
new products and culinary information flows in, 249–251
between offer, restriction, and added value of organic products in, 247–249
Meat consumption, 3–4, 83, 85
Meat, environmental impact of, 84–85
Mediterranean area, 32
addressing social sustainability, 33–34
Mediterranean diet as sustainable diet, 34–36
sustainability and sociocultural aspects, 36–43
Mediterranean culinary system, 37
Mediterranean diet, 35
heritage conservation, 36–37
as sustainable diet, 34–36
Mediterranean Strategy for Sustainable Development (MSSD), 40–41
Menu, 50–51, 54, 57–70, 59t
Metal, 101

Mexican Institute of Industrial Property (IMPI), 169
Mexico
  marketed in, 249–250
  sun and beach tourist destinations in, 240–241
Michelin stars, 70–76, 133
Migration, 241, 243, 245, 249
  configuration of new consumers in framework of, 243–244
  forced, 151–154
Mobile kitchen, 163–164
Mobility phenomena, 241–242
Mode of evaluation, 136
MOM. See Mazatlán Organic Market (MOM)
Monoculture of grains, 239–240
MSSD. See Mediterranean Strategy for Sustainable Development (MSSD)
MUCEM. See Museum of Mediterranean Cultures and Civilizations (MUCEM)
Museum of Mediterranean Cultures and Civilizations (MUCEM), 42

N
Natural biodiversity, 220
Natural food concept, 227
Natural resources, 3, 34, 37
Network concept, 186
New Nordic Cuisine, 144
Nitrates, 174
Noma considerations
  architectural or institutional context, 144
  brief contextualization of, 146–149
  cooks and customers without intermediaries, 146
  domain of wild, 140–143
  idea of test, 139–140
  labor relations, 144–146
  model of gastronomic excellence, 139
  theoretical approach and main concepts, 134–139
    characterization of third spirit of capitalism, 135t
    convention theory, 136t
  wild as an icon of indigenous, 143–144

Noma gastronomic model, 142
Noma's food, 142–143
Nonrefugee, 160
Nordic Cuisine Symposium, 144
Nordic food, 133–134
North Atlantic House, 144
Nouvelle cuisine, 49–50
Null hypothesis, 108
Nutrition, 17–18, 87–88, 215–216
Nutritional quality, 15–16
Nutritional security, 32
Nutritional transition, 83–84

O
Oilseeds, 239
Olives, 208
  oil, 207
"Orders of worth" model, 134
Organic beef, 194
Organic farming, 231–232
Organic foods, 11–12, 54–56, 229
Organic production, 187–188
Organic products, 225
Organic wastes, 174

P
Packaging process, 229
Palm leaf, 156
Patrimonialization, 137–138
Permanent migration, 244
PGI. See Protected geographical indications (PGI)
Piñonates, 208
Plastic waste, 24–25
Plastic-covered hectares, 24–25
Plátano Thai, 249–250
Political entrepreneurship, 120
Pollen, 208
Postcovid tourism systems, 96–97
Poultry consumption, 86
Poverty-stricken Hurdanos, 208
Power of food, 9–10, 151–152
Preserves, 246
Preserving traditional production methods, 187–188
Production system, 223–224, 227–228
Products, 194, 246

Index

agricultural, 36, 85
awareness, 126–128
Catalan, 130
dairy, 240–241, 246
organic, 225
traditional, 11–12
Protected geographical indications (PGI), 193
Proteins, 39
Psychic hunger, 86–87
Public foodscapes, 163–164
Public health challenge, 32
Pyrenees
in Catalonia, 186
key elements linked to territorial sustainability in, 192*t*, 195*t*
tourism brand, 190–194

Q

Q Brand. *See* Quality brands (Q Brand)
QSRs. *See* Quick-service restaurants (QSRs)
Quality brands (Q Brand), 193
Quick-service restaurants (QSRs), 96–97, 99, 103

R

Radical cosmopolitanism, 160–161
RefuEat, 153–154, 157
asylum-seekers reception centers in Berlin, 159*f*
Berlin locations visited by RefuEat staff, 158*f*
food bikes, 154
mobile menu, 156
Refugee, 160
movement, 156–157
nonemployed, 157–158
Syrian, 153
Regional Center for Integral Services for Protected Agriculture (CRESIAP), 173
René Redzepi, 133–134
Renewable energy sources, 1–2
Restaurants, 52–53, 122–123, 144.
See also Sustainable restaurants
menu, 54

safety protocols, 103
Retail, 9–10, 151–152
Robotics, 95–96
medical, 101
technologies, 96–97
Robotization, 97–98, 100–101
Robots, 98–99. *See also* Service robots
morphology of, 99–100
Routes, 172
agri-food, 170–172
gastronomic, 193
Rural culture, 171
Rural development, 171, 175

S

SADER. *See* Secretary of Agriculture and Rural Development (SADER)
Sausages, 240–241, 246
Scrap meats, 156
SDGs. *See* Sustainable development goals (SDGs)
Seasonal foods, 22–23
Seasonality, 15, 20–21, 126–128
loss of, 20–23
Secretary of Agriculture and Rural Development (SADER), 174–175
Self-made food technologies, 158–159
Service quality, 95–99
Service robots, 96, 98–99
and consumer attitudes, 101–103
in tourism, 96–101
automation in food experiences, 98–99
definition of service robots, 97–98
human-robot interactions in gastronomic experiences, 100–101
morphology of robots, 99–100
Sesame, 239
Slow food movement, 118–119
Slow Food's Snail of Approval, 122–123
Small-scale production, 187–188
Snowbirds, 245
Social actors interfere, 120–121
Social critique, 134–135
Social inclusion, 158–159
Social justice
issues, 163

Social justice (*Continued*)
  movements, 153—154
Social network, 135—136
Social processes, 249, 251
Social sustainability, 33—34
Social values, 187—188
Sofrito, 125
Soil erosion, 36
Sorghum, 239—240
Spanish Academy of Cinematic Arts and
  Sciences, 203—204
Spanish fast food experience
  data and method, 104
    consumers attitude toward AI in fast-
    food restaurants, 105t
  results, 104—109
    age and satisfaction with service
    robots, 109f
    automatization process for purchasing
    process, 106f
    human function, 107f
    Kruskal-Wallis test statistics for
    independent samples, 109t
    main statistics for Likert-type variables,
    107t
    Mann-Whitney test statistics, 108t
  service robots and consumer attitudes,
    101—103
  service robots in tourism, 96—101
    automation in food experiences,
    98—99
    definition of service robots, 97—98
    human-robot interactions in
    gastronomic experiences,
    100—101
    morphology of robots, 99—100
Spanish Global Compact Network, 189
Spirit of capitalism, 134—135
  characterization of, 135t
  model, 134
State of Sinaloa, 239
Statistical Institute of Catalonia, 190
Storage unit, 161—162
Storms, 20—21
Supermarkets, 21—22
Supply chains, 84—85
Support of migrants, 151—152

Surplus production, 31
Sustainability, 1, 15, 23—24, 54—57, 83,
  95, 123—126, 148—149, 152, 188,
  208—211
  of agri-food systems, 171
  concept, 121
  definition of sustainability criteria and
    valorization, 56t
  emergence of awareness, 49—51
  at menu of Malaysian, Indonesian, and
    Singaporean top chefs, 49
  methodology, 54—57, 59t
    sample count of chefs and restaurants,
    55t
  results, 58—75
  and sociocultural aspects, 36—43
  socioeconomic contexts, 51—53
    GDP per capita, 53f
  valorization and profiling, 58t
Sustainable agency, 127—128
Sustainable consumption, 243—244
Sustainable development, 33, 91, 185,
  189—190
  of businesses, 110
  of gastronomic tourism, 191
Sustainable development goals (SDGs),
  2—3, 185
Sustainable diets, 84
  China and Brazil, 86—89
  intergenerational discussion, 89—91
  meat, environmental impact of,
    84—85
Sustainable dishes, 209—210
Sustainable food, 118—119
Sustainable gastronomic tourism, 188—189
Sustainable restaurants, 119—121
  Barcelona as cosmopolitan city,
    128—130
  Catalan cuisine, 123—126
  methodology, 121—123
  seasonality, product awareness, and
    cooperation with local producers,
    126—128
Sustainable rural development
  agri-food routes and, 170—172
  Yahualica chili route and, 178—181
Sweet pastries, 208

Index

T

Temporary migration, 244
Tentation/s program, 207
Territorial sustainability, key elements linked to, 192t
Tomato, 239, 249–250
Topography, 201
Tortas ahogadas, 175–176
Tourism, 205, 241, 243, 249
  configuration of new consumers in framework of, 243–244
  model, 209–210
  systems, 96–97
  value chain, 187–188
Traditional agricultural landscapes, 36
Traditional circuits, 229
Traditional farmers, 220
Traditional Hurdano recipes, 207–209
Traditional products, 11–12
Transforming Traditional Agriculture (Schultz), 217–219
Transnational food chains, 83
Typology of orders, 136

U

UNESCO, 36–37
United Nations, 191
United Nations General Assembly, 185
Urbanization process, 52, 85

V

Vaccinations, 26–27
Valorization, 10
  definition of sustainability criteria and, 56t
  and profiling, 58t
Vegetables, 18, 20, 23–24, 24f, 129, 239
Vegetalization, 6–7, 57–70, 76–77
Vitamin-rich diets, 87–88
Vitamins, 39

W

Warehouses, 25–26
Waste management, 6–7, 50–51
Water footprint, 35–36, 84
Water resources, 36
Weaving system, 225–226
Wild mushrooms, 208
  from Cerdanya valley, 25f
Winter birds, 245
WLA. *See* Women's Land Army (WLA)
Women's Land Army (WLA), 19
World Heritage Site, 221–222
World Tourism Organization, The, 189

Y

Yahualica chilli (*Capsicum annuum*), 169, 172
  challenges in production of, 173–174
  gastronomic and food uses of, 175–176
  route, 178–179
    and sustainable rural development, 178–181
  tourist and gastronomic activations of, 176–177

Z

Zero miles, 209–210

Printed in the United States
by Baker & Taylor Publisher Services